CRC
STANDARD PROBABILITY AND STATISTICS TABLES AND FORMULAE

STUDENT EDITION

CRC
STANDARD PROBABILITY AND STATISTICS TABLES AND FORMULAE

STUDENT EDITION

STEPHEN KOKOSKA
Bloomsburg University
Bloomsburg, Pennsylvania

DANIEL ZWILLINGER
Rensselaer Polytechnic Institute
Troy, New York

CHAPMAN & HALL/CRC
Boca Raton London New York Washington, D.C.

Much of this book originally appeared in D. Zwillinger and S. Kokoska, *Standard Probability and Statistics Tables and Formulae*, Chapman & Hall/CRC, 2000. Reprinted courtesy of Chapman & Hall/CRC.

Library of Congress Cataloging-in-Publication Data

Catalog record is available from the Library of Congress.

This book contains information obtained from authentic and highly regarded sources. Reprinted material is quoted with permission, and sources are indicated. A wide variety of references are listed. Reasonable efforts have been made to publish reliable data and information, but the author and the publisher cannot assume responsibility for the validity of all materials or for the consequences of their use.

Neither this book nor any part may be reproduced or transmitted in any form or by any means, electronic or mechanical, including photocopying, microfilming, and recording, or by any information storage or retrieval system, without prior permission in writing from the publisher.

The consent of CRC Press LLC does not extend to copying for general distribution, for promotion, for creating new works, or for resale. Specific permission must be obtained in writing from CRC Press LLC for such copying.

Direct all inquiries to CRC Press LLC, 2000 N.W. Corporate Blvd., Boca Raton, Florida 33431.

Trademark Notice: Product or corporate names may be trademarks or registered trademarks, and are used only for identification and explanation, without intent to infringe.

© 2000 by Chapman & Hall/CRC

No claim to original U.S. Government works
International Standard Book Number 0-8493-0026-6
Printed in the United States of America 1 2 3 4 5 6 7 8 9 0
Printed on acid-free paper

Preface

It has long been the established policy of CRC Press to publish, in handbook form, the most up-to-date, authoritative, logically arranged, and readily usable reference material available. More recently, CRC Press has published student editions of the more extensive reference texts.

The purpose of this book is to provide a modern set of tables and a comprehensive list of definitions, concepts, theorems, and formulae for students enrolled in introductory probability and statistics classes. While the numbers in these tables have not changed since they were first computed (in some cases, several hundred years ago), the presentation format here is modernized. In addition, nearly all table values have been recomputed to ensure accuracy.

It has become less important to memorize formulas but essential for students to learn which probability or statistical formula or technique to use in order to solve a given problem. This reference is designed to provide a concise understandable description of introductory probability and statistics procedures. We anticipate both students and instructors will add to this material. We hope this supplement will replace text-specific probability and statistics formula cards and that students will refer to this supplement to verify techniques and to complete assignments and exams.

In line with the established policy of CRC Press, this *Student Edition Handbook* will be kept as current and timely as is possible. Revisions and anticipated uses of newer materials and tables will be introduced as the need arises. Suggestions for the inclusion of new material in subsequent editions and comments concerning the accuracy of stated information are welcomed.

If any errata are discovered for this book, they will be posted to
http://vesta.bloomu.edu/~skokoska/prast/errata.

Many individuals have helped in the preparation of this manuscript. We are especially grateful to our families who have remained lighthearted and cheerful throughout the process. A special thanks to to Joan, Mark, and Jen (for enforcing the coffee-break rule), and to Janet and Kent.

Stephen Kokoska   Daniel Zwillinger
skokoska@planetx.bloomu.edu   zwillinger@alum.mit.edu

Contents

1		**Introduction**	**1**
	1.1	Data sets	1
2		**Summarizing Data**	**3**
	2.1	Tabular and graphical procedures	3
	2.2	Numerical summary measures	7
3		**Probability**	**19**
	3.1	Algebra of sets	19
	3.2	Combinatorial methods	19
	3.3	Probability	23
	3.4	Random variables	27
	3.5	Mathematical expectation	29
	3.6	Multivariate distributions	32
	3.7	Inequalities	38
4		**Functions of Random Variables**	**39**
	4.1	Finding the probability distribution	39
	4.2	Sums of random variables	43
	4.3	Sampling distributions	44
	4.4	Finite population	46
	4.5	Theorems	47
	4.6	Order statistics	48
	4.7	Range and studentized range	53
5		**Discrete Probability Distributions**	**59**
	5.1	Bernoulli distribution	59

5.2	Binomial distribution	60
5.3	Geometric distribution	66
5.4	Hypergeometric distribution	68
5.5	Multinomial distribution	70
5.6	Negative binomial distribution	70
5.7	Poisson distribution	72
5.8	Rectangular (discrete uniform) distribution	75

6 Continuous Probability Distributions — 77

6.1	Cauchy distribution	79
6.2	chi-square distribution	80
6.3	Erlang distribution	87
6.4	Exponential distribution	88
6.5	F distribution	89
6.6	Gamma distribution	96
6.7	Lognormal distribution	97
6.8	Normal distribution	99
6.9	Normal distribution: multivariate	101
6.10	Pareto distribution	102
6.11	Rayleigh distribution	103
6.12	t distribution	104
6.13	Triangular distribution	107
6.14	Uniform distribution	108
6.15	Weibull distribution	109
6.16	Relationships among distributions	110

7 Standard Normal Distribution — 115

7.1	Density function and related functions	115
7.2	Critical values	125
7.3	Tolerance factors for normal distributions	125
7.4	Operating characteristic curves	128
7.5	Multivariate normal distribution	131
7.6	Distribution of the correlation coefficient	131

8 Estimation — 135

8.1	Definitions	135
8.2	Cramér–Rao inequality	136

8.3	Theorems	137
8.4	The method of moments	138
8.5	The likelihood function	138
8.6	The method of maximum likelihood	138
8.7	Invariance property of MLEs	139
8.8	Different estimators	139

9 Confidence Intervals   141

9.1	Definitions	141
9.2	Common critical values	141
9.3	Sample size calculations	142
9.4	Summary of common confidence intervals	143
9.5	Other tests	144
9.6	Finite population correction factor	146

10 Hypothesis Testing   147

10.1	Introduction	147
10.2	The Neyman–Pearson lemma	151
10.3	Likelihood ratio tests	151
10.4	Goodness of fit test	151
10.5	Contingency tables	153
10.6	Significance test in 2×2 contingency tables	154
10.7	Critical values for testing outliers	155

11 Regression Analysis   157

11.1	Simple linear regression	157
11.2	Multiple linear regression	164

12 Nonparametric Statistics   165

12.1	Friedman test for randomized block design	165
12.2	Kendall's rank correlation coefficient	165
12.3	Kolmogorov–Smirnoff tests	167
12.4	Kruskal–Wallis test	173
12.5	The runs test	175
12.6	The sign test	185
12.7	Spearman's rank correlation coefficient	186
12.8	Wilcoxon matched-pairs signed-ranks test	191

| | 12.9 | Wilcoxon rank–sum (Mann–Whitney) test . . . . 192 |
| | 12.10 | Wilcoxon signed-rank test . . . . 200 |

13 Miscellaneous topics   201
    13.1   Ceiling and floor functions 201
    13.2   Error functions 201
    13.3   Exponential function 202
    13.4   Factorials and Pochhammer's symbol 203
    13.5   Gamma function 204
    13.6   Hypergeometric functions 206
    13.7   Logarithmic functions 207
    13.8   Sums of powers of integers 208

       Notation   **212**

       Index   **217**

CHAPTER 1

Introduction

1.1 DATA SETS

This section contains several data sets used in examples throughout this book. With these, a user can check a local statistics program by verifying that it returns the same values as given in this book. These data sets may be obtained from http://vesta.bloomu.edu/~skokoska/prast/data.

Ticket data: Forty random speeding tickets were selected from the courthouse records in Columbia County. The speed indicated on each ticket is given in the table below.

58	72	64	65	67	92	55	51	69	73
64	59	65	55	75	56	89	60	84	68
74	67	55	68	74	43	67	71	72	66
62	63	83	64	51	63	49	78	65	75

Swimming pool data: Water samples from 35 randomly selected pools in Beverly Hills were tested for acidity. The following table lists the pH for each sample.

6.4	6.6	6.2	7.2	6.2	8.1	7.0
7.0	5.9	5.7	7.0	7.4	6.5	6.8
7.0	7.0	6.0	6.3	5.6	6.3	5.8
5.9	7.2	7.3	7.7	6.8	5.2	5.2
6.4	6.3	6.2	7.5	6.7	6.4	7.8

Soda pop data: A new soda machine placed in the Mathematics Building on campus recorded the following sales data for one week in April.

Soda	Number of cans
Pepsi	72
Wild Cherry Pepsi	60
Diet Pepsi	85
Seven Up	54
Mountain Dew	32
Lipton Ice Tea	64

CHAPTER 2

Summarizing Data

Numerical descriptive statistics and graphical techniques may be used to summarize information about central tendency and/or variability.

2.1 TABULAR AND GRAPHICAL PROCEDURES

2.1.1 Stem-and-leaf plot

A stem-and-leaf plot is a a graphical summary used to describe a set of observations (as symmetric, skewed, etc.). Each observation is displayed on the graph and should have at least two digits. Split each observation (at the same point) into a stem (one or more of the leading digit(s)) and a leaf (remaining digits). Select the split point so that there are 5–20 total stems. List the stems in a column to the left, and write each leaf in the corresponding stem row.

Example 2.1: Construct a stem-and-leaf plot for the Ticket Data (page 1).

Solution:

Stem	Leaf
4	3 9
5	1 1 5 5 5 6 8 9
6	0 2 3 3 4 4 4 5 5 5 6 7 7 7 8 8 9
7	1 2 2 3 4 4 5 5 8
8	3 4 9
9	2

Stem = 10, Leaf = 1

Figure 2.1: Stem-and-leaf plot for Ticket Data.

2.1.2 Frequency distribution

A frequency distribution is a tabular method for summarizing continuous or discrete numerical data or categorical data.

(1) Partition the measurement axis into 5–20 (usually equal) reasonable subintervals called classes, or class intervals. Thus, each observation falls into exactly one class.

(2) Record, or tally, the number of observations in each class, called the frequency of each class.

(3) Compute the proportion of observations in each class, called the relative frequency.

(4) Compute the proportion of observations in each class *and* all preceding classes, called the cumulative relative frequency.

Example 2.2: Construct a frequency distribution for the Ticket Data (page 1).

Solution:

(S1) Determine the classes. It seems reasonable to use 40 to less than 50, 50 to less than 60, ..., 90 to less than 100.

Note: For continuous data, one end of each class must be open. This ensures that each observation will fall into only one class. The open end of each class may be either the left or right, but should be consistent.

(S2) Record the number of observations in each class.

(S3) Compute the relative frequency and cumulative relative frequency for each class.

(S4) The resulting frequency distribution is in Figure 2.2.

Class	Frequency	Relative frequency	Cumulative relative frequency
$[40, 50)$	2	0.050	0.050
$[50, 60)$	8	0.200	0.250
$[60, 70)$	17	0.425	0.625
$[70, 80)$	9	0.225	0.900
$[80, 90)$	3	0.075	0.975
$[90, 100)$	1	0.025	1.000

Figure 2.2: Frequency distribution for Ticket Data.

2.1.3 Histogram

A histogram is a graphical representation of a frequency distribution. A (relative) frequency histogram is a plot of (relative) frequency versus class interval. Rectangles are constructed over each class with height proportional (usually equal) to the class (relative) frequency. A frequency and relative frequency histogram have the same shape, but different scales on the vertical axis.

Example 2.3: Construct a frequency histogram for the Ticket Data (page 1).

Solution:

(S1) Using the frequency distribution in Figure 2.2, construct rectangles above each class, with height equal to class frequency.

(S2) The resulting histogram is in Figure 2.3.

2.1. TABULAR AND GRAPHICAL PROCEDURES

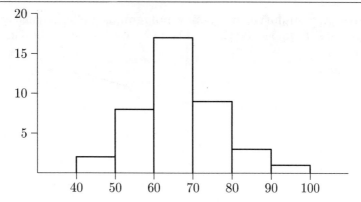

Figure 2.3: Frequency histogram for Ticket Data.

Note: A probability histogram is constructed so that the area of each rectangle equals the relative frequency. If the class widths are unequal, this histogram presents a more accurate description of the distribution.

2.1.4 Frequency polygons

A frequency polygon is a line plot of points with x coordinate being class midpoint and y coordinate being class frequency. Often the graph extends to an additional empty class on both ends. The relative frequency may be used in place of frequency.

Example 2.4: Construct a frequency polygon for the Ticket Data (page 1).

Solution:

(S1) Using the frequency distribution in Figure 2.2, plot each point and connect the graph.

(S2) The resulting frequency polygon is in Figure 2.4.

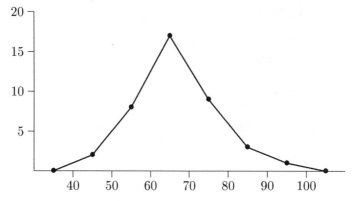

Figure 2.4: Frequency polygon for Ticket Data.

An **ogive**, or **cumulative frequency polygon**, is a plot of cumulative frequency versus the upper class limit. Figure 2.5 is an ogive for the Ticket Data (page 1).

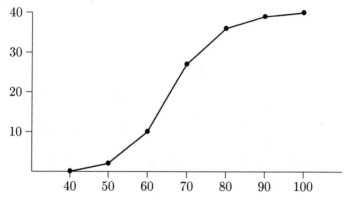

Figure 2.5: Ogive for Ticket Data.

Another type of frequency polygon is a *more-than* cumulative frequency polygon. For each class this plots the number of observations in that class and every class above versus the lower class limit.

A **bar chart** is often used to graphically summarize discrete or categorical data. A rectangle is drawn over each bin with height proportional to frequency. The chart may be drawn with horizontal rectangles, in three dimensions, and may be used to compare two or more sets of observations. Figure 2.6 is a bar chart for the Soda Pop Data (page 1).

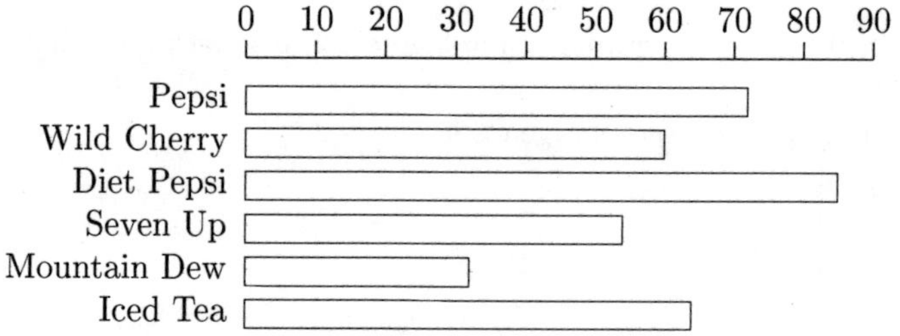

Figure 2.6: Bar chart for Soda Pop Data.

A **pie chart** is used to illustrate parts of the total. A circle is divided into slices proportional to the bin frequency. Figure 2.7 is a pie chart for the Soda Pop Data (page 1).

2.2. NUMERICAL SUMMARY MEASURES 7

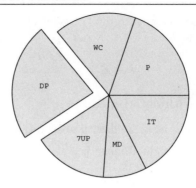

Figure 2.7: Pie chart for Soda Pop Data.

2.2 NUMERICAL SUMMARY MEASURES

The following conventions will be used in the definitions and formulas in this section.

(C1) *Ungrouped data*: Let $x_1, x_2, x_3, \ldots, x_n$ be a set of observations.

(C2) *Grouped data*: Let $x_1, x_2, x_3, \ldots, x_k$ be a set of class marks from a frequency distribution, or a representative set of observations, with corresponding frequencies $f_1, f_2, f_3, \ldots, f_k$. The total number of observations is $n = \sum_{i=1}^{k} f_i$. Let c denote the (constant) width of each bin and x_o one of the class marks selected to be the *computing origin*. Each class mark, x_i, may be coded by $u_i = (x_i - x_o)/c$. Each u_i will be an integer and the bin mark taken as the computing origin will be coded as a 0.

2.2.1 (Arithmetic) mean

The (arithmetic) mean of a set of observations is the sum of the observations divided by the total number of observations.

(1) Ungrouped data:
$$\bar{x} = \frac{1}{n} \sum_{i=1}^{n} x_i = \frac{x_1 + x_2 + x_3 + \cdots + x_n}{n} \tag{2.1}$$

(2) Grouped data:
$$\bar{x} = \frac{1}{n} \sum_{i=1}^{k} f_i x_i = \frac{f_1 x_1 + f_2 x_2 + f_3 x_3 + \cdots + f_n x_n}{n} \tag{2.2}$$

(3) Coded data:
$$\bar{x} = x_o + c \cdot \frac{\sum_{i=1}^{k} f_i u_i}{n} \qquad (2.3)$$

2.2.2 Weighted (arithmetic) mean

Let $w_i \geq 0$ be the weight associated with observation x_i. The *total weight* is given by $\sum_{i=1}^{n} w_i$ and the weighted mean is

$$\bar{x}_w = \frac{\sum_{i=1}^{n} w_i x_i}{\sum_{i=1}^{n} w_i} = \frac{w_1 x_1 + w_2 x_2 + w_3 x_3 + \cdots + w_n x_n}{w_1 + w_2 + w_3 + \cdots + w_n}. \qquad (2.4)$$

2.2.3 Geometric mean

For ungrouped data such that $x_i > 0$, the geometric mean is the nth root of the product of the observations:

$$\text{GM} = \sqrt[n]{x_1 \cdot x_2 \cdot x_3 \cdots x_n}. \qquad (2.5)$$

In logarithmic form:

$$\log(\text{GM}) = \frac{1}{n} \sum_{i=1}^{n} \log x_i = \frac{\log x_1 + \log x_2 + \log x_3 + \cdots + \log x_n}{n}. \qquad (2.6)$$

For grouped data with each class mark $x_i > 0$:

$$\text{GM} = \sqrt[n]{x_1^{f_1} \cdot x_2^{f_2} \cdot x_3^{f_3} \cdots x_k^{f_k}}. \qquad (2.7)$$

In logarithmic form:

$$\log(\text{GM}) = \frac{1}{n} \sum_{i=1}^{k} f_i \log(x_i) \qquad (2.8)$$

$$= \frac{f_1 \log(x_1) + f_2 \log(x_2) + f_3 \log(x_3) + \cdots + f_k \log(x_k)}{n}.$$

2.2.4 Harmonic mean

For ungrouped data the harmonic mean is given by

$$\text{HM} = \frac{n}{\sum_{i=1}^{n} \frac{1}{x_i}} = \frac{n}{\frac{1}{x_1} + \frac{1}{x_2} + \frac{1}{x_3} + \cdots + \frac{1}{x_n}}. \qquad (2.9)$$

2.2. NUMERICAL SUMMARY MEASURES

For grouped data:

$$\text{HM} = \frac{n}{\sum_{i=1}^{k} \frac{f_i}{x_i}} = \frac{n}{\frac{f_1}{x_1} + \frac{f_2}{x_2} + \frac{f_3}{x_3} + \cdots + \frac{f_k}{x_k}}. \tag{2.10}$$

Note: The equation involving the arithmetic, geometric, and harmonic mean is

$$\text{HM} \leq \text{GM} \leq \bar{x}. \tag{2.11}$$

Equality holds if all n observations are equal.

2.2.5 Mode

For ungrouped data, the mode, M_o, is the value that occurs most often, or with the greatest frequency. A mode may not exist, for example, if all observations occur with the same frequency. If the mode does exist, it may not be unique, for example, if two observations occur with the greatest frequency.

For grouped data, select the class containing the largest frequency, called the modal class. Let L be the lower boundary of the modal class, d_L the difference in frequencies between the modal class and the class immediately below, and d_H the difference in frequencies between the modal class and the class immediately above. The mode may be approximated by

$$M_o \approx L + c \cdot \frac{d_L}{d_L + d_H}. \tag{2.12}$$

2.2.6 Median

The median, $\tilde{x}$, is another measure of central tendency, resistant to outliers. For ungrouped data, arrange the observations in order from smallest to largest. If n is odd, the median is the middle value. If n is even, the median is the mean of the two middle values.

For grouped data, select the class containing the median (median class). Let L be the lower boundary of the median class, f_m the frequency of the median class, and CF the sum of frequencies for all classes below the median class (a cumulative frequency). The median may be approximated by

$$\tilde{x} \approx L + c \cdot \frac{\frac{n}{2} - \text{CF}}{f_m}. \tag{2.13}$$

Note: If $\bar{x} > \tilde{x}$ the distribution is positively skewed. If $\bar{x} < \tilde{x}$ the distribution is negatively skewed. If $\bar{x} \approx \tilde{x}$ the distribution is approximately symmetric.

2.2.7 $p\%$ trimmed mean

A trimmed mean is a measure of central tendency and a compromise between a mean and a median. The mean is more sensitive to outliers, and the median is less sensitive to outliers. Order the observations from smallest to largest.

Delete the smallest $p\%$ *and* the largest $p\%$ of the observations. The $p\%$ trimmed mean, $\bar{x}_{\text{tr}(p)}$, is the arithmetic mean of the remaining observations.

Note: If $p\%$ of n (observations) is not an integer, several (computer) algorithms exist for interpolating at each end of the distribution and for determining $\bar{x}_{\text{tr}(p)}$.

Example 2.5: Using the Swimming Pool data (page 1) find the mean, median, and mode. Compute the geometric mean and the harmonic mean, and verify the relationship between these three measures.

Solution:

(S1) $\bar{x} = \dfrac{1}{35}(6.4 + 6.6 + 6.2 + \cdots + 7.8) = 6.5886$

(S2) $\tilde{x} = 6.5$, the middle values when the observations are arranged in order from smallest to largest.

(S3) $M_o = 7.0$, the observation that occurs most often.

(S4) $GM = \sqrt[35]{(6.4)(6.6)(6.2)\cdots(7.8)} = 6.5513$

(S5) $HM = \dfrac{35}{(1/6.4) + (1/6.6) + (1/6.2) + \cdots + (1/7.8)} = 6.5137$

(S6) To verify the inequality: $\underbrace{6.5137}_{HM} \leq \underbrace{6.5513}_{GM} \leq \underbrace{6.5886}_{\bar{x}}$

2.2.8 Quartiles

Quartiles split the data into four parts. For ungrouped data, arrange the observations in order from smallest to largest.

(1) The second quartile is the median: $Q_2 = \tilde{x}$.

(2) If n is even:

   The first quartile, Q_1, is the median of the smallest $n/2$ observations; and the third quartile, Q_3, is the median of the largest $n/2$ observations.

(3) If n is odd:

   The first quartile, Q_1, is the median of the smallest $(n+1)/2$ observations; and the third quartile, Q_3, is the median of the largest $(n+1)/2$ observations.

For grouped data, the quartiles are computed by applying equation (2.13) for the median. Compute the following:

L_1 = the lower boundary of the class containing Q_1.

L_3 = the lower boundary of the class containing Q_3.

f_1 = the frequency of the class containing the first quartile.

f_3 = the frequency of the class containing the third quartile.

CF_1 = cumulative frequency for classes below the one containing Q_1.

CF_3 = cumulative frequency for classes below the one containing Q_3.

2.2. NUMERICAL SUMMARY MEASURES

The (approximate) quartiles are given by

$$Q_1 = L_1 + c \cdot \frac{\frac{n}{4} - CF_1}{f_1} \qquad Q_3 = L_3 + c \cdot \frac{\frac{3n}{4} - CF_3}{f_3}. \qquad (2.14)$$

2.2.9 Deciles

Deciles split the data into 10 parts.

(1) For ungrouped data, arrange the observations in order from smallest to largest. The ith decile, D_i (for $i = 1, 2, \ldots, 9$), is the $i(n+1)/10^{th}$ observation. It may be necessary to interpolate between successive values.

(2) For grouped data, apply equation (2.13) (as in equation (2.14)) for the median to find the approximate deciles. D_i is in the class containing the $in/10^{th}$ largest observation.

2.2.10 Percentiles

Percentiles split the data into 100 parts.

(1) For ungrouped data, arrange the observations in order from smallest to largest. The ith percentile, P_i (for $i = 1, 2, \ldots, 99$), is the $i(n+1)/100^{th}$ observation. In some cases it may be necessary to interpolate between successive values.

(2) For grouped data, apply equation (2.13) (as in equation (2.14)) for the median to find the approximate percentiles. P_i is in the class containing the $in/100^{th}$ largest observation.

2.2.11 Mean deviation

The mean deviation is a measure of variability based on the absolute value of the deviations about the mean or median.

(1) For ungrouped data:

$$\text{MD} = \frac{1}{n} \sum_{i=1}^{n} |x_i - \bar{x}| \quad \text{or} \quad \text{MD} = \frac{1}{n} \sum_{i=1}^{n} |x_i - \tilde{x}|. \qquad (2.15)$$

(2) For grouped data:

$$\text{MD} = \frac{1}{n} \sum_{i=1}^{k} f_i |x_i - \bar{x}| \quad \text{or} \quad \text{MD} = \frac{1}{n} \sum_{i=1}^{k} f_i |x_i - \tilde{x}|. \qquad (2.16)$$

2.2.12 Variance

The variance is a measure of variability based on the squared deviations about the mean.

(1) For ungrouped data:

$$s^2 = \frac{1}{n-1} \sum_{i=1}^{n} (x_i - \overline{x})^2. \qquad (2.17)$$

The computational formula for s^2:

$$s^2 = \frac{1}{n-1} \left[\sum_{i=1}^{n} x_i^2 - \frac{1}{n} \left(\sum_{i=1}^{n} x_i \right)^2 \right] = \frac{1}{n-1} \left(\sum_{i=1}^{n} x_i^2 - n\overline{x}^2 \right). \qquad (2.18)$$

(2) For grouped data:

$$s^2 = \frac{1}{n-1} \sum_{i=1}^{k} f_i (x_i - \overline{x})^2. \qquad (2.19)$$

The computational formula for s^2:

$$s^2 = \frac{1}{n-1} \left[\sum_{i=1}^{k} f_i x_i^2 - \frac{1}{n} \left(\sum_{i=1}^{k} f_i x_i \right)^2 \right]$$
$$= \frac{1}{n-1} \left(\sum_{i=1}^{k} f_i x_i^2 - n\overline{x}^2 \right). \qquad (2.20)$$

(3) For coded data:

$$s^2 = \frac{c}{n-1} \left[\sum_{i=1}^{k} f_i u_i^2 - \frac{1}{n} \left(\sum_{i=1}^{k} f_i u_i \right)^2 \right]. \qquad (2.21)$$

2.2.13 Standard deviation

The *standard deviation* is the positive square root of the variance: $s = \sqrt{s^2}$. The *probable error* is 0.6745 times the standard deviation.

2.2.14 Standard errors

The standard error of a statistic is the standard deviation of the sampling distribution of that statistic. The standard error of a statistic is often designated by σ with a subscript indicating the statistic.

2.2.14.1 Standard error of the mean

The standard error of the mean is used in hypothesis testing and is an indication of the accuracy of the estimate $\overline{x}$.

$$\text{SEM} = s/\sqrt{n}. \qquad (2.22)$$

2.2. NUMERICAL SUMMARY MEASURES

2.2.15 Root mean square

(1) For ungrouped data:

$$\text{RMS} = \left(\frac{1}{n}\sum_{i=1}^{n} x_i^2\right)^{1/2}. \tag{2.23}$$

(2) For grouped data:

$$\text{RMS} = \left(\frac{1}{n}\sum_{i=1}^{k} f_i x_i^2\right)^{1/2}. \tag{2.24}$$

2.2.16 Range

The range is the difference between the largest and smallest values.

$$R = \max\{x_1, x_2, \ldots, x_n\} - \min\{x_1, x_2, \ldots, x_n\} = x_{(n)} - x_{(1)}. \tag{2.25}$$

2.2.17 Interquartile range

The interquartile range, or fourth spread, is the difference between the third and first quartile.

$$\text{IQR} = Q_3 - Q_1. \tag{2.26}$$

2.2.18 Quartile deviation

The quartile deviation, or semi-interquartile range, is half the interquartile range.

$$\text{QD} = \frac{Q_3 - Q_1}{2}. \tag{2.27}$$

2.2.19 Box plots

Box plots, also known as quantile plots, are graphics which display the center portions of the data and some information about the range of the data. There are a number of variations and a box plot may be drawn with either a horizontal or vertical scale. The *inner* and *outer fences* are used in constructing a box plot and are markers used in identifying mild and extreme outliers.

$$\begin{aligned}\text{Inner Fences:} \quad & Q_1 - 1.5 \cdot \text{IQR}, \quad Q_1 + 1.5 \cdot \text{IQR} \\ \text{Outer Fences:} \quad & Q_3 - 3 \cdot \text{IQR}, \quad Q_3 + 3 \cdot \text{IQR}\end{aligned} \tag{2.28}$$

A general description:

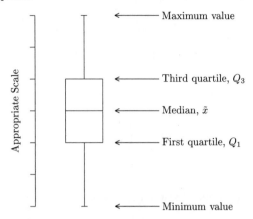

Multiple box plots on the same measurement axis may be used to compare the center and spread of distributions. Figure 2.8 presents box plots for randomly selected August residential electricity bills for three different parts of the country.

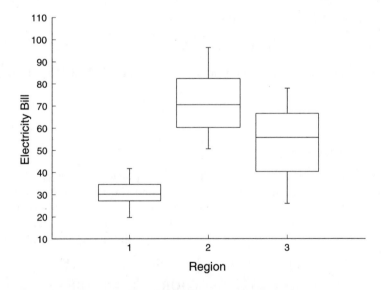

Figure 2.8: Example of multiple box plots.

2.2. NUMERICAL SUMMARY MEASURES

2.2.20 Coefficient of variation

The coefficient of variation is a measure of relative variability. Reported as percentage it is defined as:

$$\mathrm{CV} = 100\,\frac{s}{\bar{x}}. \tag{2.29}$$

2.2.21 Coefficient of quartile variation

The coefficient of quartile variation is a measure of variability.

$$\mathrm{CQV} = 100\,\frac{Q_3 - Q_1}{Q_3 + Q_1}. \tag{2.30}$$

2.2.22 Z score

The z score, or standard score, associated with an observation is a measure of relative standing.

$$z = \frac{x_i - \bar{x}}{s} \tag{2.31}$$

2.2.23 Moments

Moments are used to characterize a set of observations.

(1) For ungrouped data:
The rth moment about the origin:

$$m'_r = \frac{1}{n} \sum_{i=1}^{n} x_i^r. \tag{2.32}$$

The rth moment about the mean $\bar{x}$:

$$m_r = \frac{1}{n} \sum_{i=1}^{n} (x_i - \bar{x})^r = \sum_{j=0}^{r} \binom{r}{j} (-1)^j m'_{r-j} \bar{x}^j. \tag{2.33}$$

(2) For grouped data:
The rth moment about the origin:

$$m'_r = \frac{1}{n} \sum_{i=1}^{k} f_i x_i^r. \tag{2.34}$$

The rth moment about the mean $\bar{x}$:

$$m_r = \frac{1}{n} \sum_{i=1}^{k} f_i (x_i - \bar{x})^r = \sum_{j=0}^{r} \binom{r}{j} (-1)^j m'_{r-j} \bar{x}^j. \tag{2.35}$$

(3) For coded data:

$$m'_r = \frac{c^r}{n} \sum_{i=1}^{n} f_i u_i^r. \tag{2.36}$$

2.2.24 Measures of skewness

The following descriptive statistics measure the lack of symmetry. Larger values (in magnitude) indicate more skewness in the distribution of observations.

2.2.24.1 Coefficient of skewness

$$g_1 = \frac{m_3}{m_2^{3/2}} \tag{2.37}$$

2.2.24.2 Coefficient of momental skewness

$$\frac{g_1}{2} = \frac{m_3}{2m_2^{3/2}} \tag{2.38}$$

2.2.24.3 Pearson's first coefficient of skewness

$$S_{k_1} = \frac{3(\bar{x} - M_o)}{s} \tag{2.39}$$

2.2.24.4 Pearson's second moment of skewness

$$S_{k_2} = \frac{3(\bar{x} - \tilde{x})}{s} \tag{2.40}$$

2.2.24.5 Quartile coefficient of skewness

$$S_{k_q} = \frac{Q_3 - 2\tilde{x} + Q_1}{Q_3 - Q_1} \tag{2.41}$$

Example 2.6: Using the Swimming Pool data (page 1) find the coefficient of skewness, coefficient of momental skewness, Pearson's first coefficient of skewness, Pearson's second moment of skewness, and the quartile coefficient of skewness.

Solution:

(S1) $\bar{x} = 6.589$, $\tilde{x} = 6.5$, $s = 0.708$, $Q_1 = 6.2$, $Q_3 = 7.0$, $M_o = 7.0$

(S2) $m_2 = \frac{1}{n}\sum_{i=1}^{35}(x_i - \bar{x})^2 = 0.4867 \quad m_3 = \frac{1}{n}\sum_{i=1}^{35}(x_i - \bar{x})^3 = 0.0126$

(S3) $g_1 = 0.0126/(0.4867)^{3/2} = 0.0371$, $\quad g_1/2 = 0.0372/2 = 0.0186$

(S4) $S_{k_1} = \frac{3(6.589 - 7)}{0.708} = -1.7415$, $\quad S_{k_2} = \frac{3(6.589 - 6.5)}{0.708} = 0.3771$

(S5) $S_{k_q} = \frac{7.0 - 2(6.5) + 6.2}{7.0 - 6.2} = 0.25$

2.2.25 Measures of kurtosis

The following statistics describe the extent of the peak in a distribution. Smaller values (in magnitude) indicate a flatter, more uniform distribution.

2.2. NUMERICAL SUMMARY MEASURES

2.2.25.1 Coefficient of kurtosis

$$g_2 = \frac{m_4}{m_2^2} \qquad (2.42)$$

2.2.25.2 Coefficient of excess kurtosis

$$g_2 - 3 = \frac{m_4}{m_2^2} - 3 \qquad (2.43)$$

2.2.26 Data transformations

Suppose $y_i = ax_i + b$ for $i = 1, 2, \ldots, n$. The following summary statistics for the distribution of y's are related to summary statistics for the distribution of x's.

$$\bar{y} = a\bar{x} + b, \qquad s_y^2 = a^2 s_x^2, \qquad s_y = |a|s_x \qquad (2.44)$$

2.2.27 Sheppard's corrections for grouping

For grouped data, suppose every class interval has width c. If both tails of the distribution are very flat and close to the measurement axis, the grouped data approximation to the sample variance may be improved by using Sheppard's correction, $-c^2/12$:

$$\text{corrected variance} = \text{grouped data variance} - \frac{c^2}{12} \qquad (2.45)$$

There are similar corrected sample moments, denoted m'_{r_c} and m_{r_c}:

$$\begin{aligned}
m'_{1_c} &= m'_1 & m_{1_c} &= m_1 \\
m'_{2_c} &= m'_2 - \frac{c^2}{12} & m_{2_c} &= m_2 - \frac{c^2}{12} \\
m'_{3_c} &= m'_3 - \frac{c^2}{4} m'_1 & m_{3_c} &= m_3 \\
m'_{4_c} &= m'_4 - \frac{c^2}{2} m'_1 + \frac{7c^2}{240} & m_{4_c} &= m_4 - \frac{c^2}{2} m_2 + \frac{7c^2}{240}
\end{aligned} \qquad (2.46)$$

Example 2.7: Consider the *grouped* Ticket Data (page 1) as presented in the frequency distribution in Example 2.2 (on page 4). Find the corrected sample variance and corrected sample moments.

Solution:

(S1) $\bar{x} = 66.5$, $s^2 = 115.64$ (for grouped data), $c = 10$

(S2) corrected variance $= 115.64 - (10^2/12) = 107.31$

(S3)

r	m'_r	m'_{r_c}	m_r	m_{r_c}
1	66.5	66.5	0.0	0.0
2	4535.0	4526.7	112.8	104.4
3	316962.5	315300.0	389.3	389.3
4	22692125.0	22688802.9	40637.3	35002.7

CHAPTER 3
Probability

3.1 ALGEBRA OF SETS

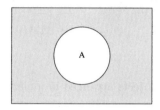

Figure 3.1: Shaded region = A'.

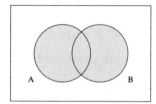

Figure 3.2: Shaded region = $A \cup B$.

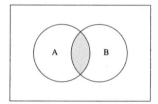

Figure 3.3: Shaded region = $A \cap B$.

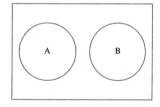

Figure 3.4: Mutually exclusive sets.

3.2 COMBINATORIAL METHODS

In an equally likely outcome experiment, computing the probability of an event involves counting. The following techniques are useful for determining the number of outcomes in an event and/or the sample space.

3.2.1 The product rule for ordered pairs

If the first element of an ordered pair can be selected in n_1 ways, and for each of these n_1 ways the second element of the pair can be selected in n_2 ways, then the number of possible pairs is $n_1 n_2$.

3.2.2 The generalized product rule for k-tuples

Suppose a sample space, or set, consists of ordered collections of k-tuples. If there are n_1 choices for the first element, and for each choice of the first element there are n_2 choices for the second element, ..., and for each of the first $k-1$ elements there are n_k choices for the kth element, then there are $n_1 n_2 \cdots n_k$ possible k-tuples.

3.2.3 Permutations

The number of permutations of n distinct objects taken k at a time is

$$P(n,k) = \frac{n!}{(n-k)!}. \tag{3.1}$$

A table of values is on page 210.

3.2.4 Circular permutations

The number of permutations of n distinct objects arranged in a circle is $(n-1)!$.

3.2.5 Combinations (binomial coefficients)

The binomial coefficient $\binom{n}{k}$ is the number of combinations of n distinct objects taken k at a time without regard to order:

$$C(n,k) = \binom{n}{k} = \frac{n!}{k!(n-k)!} = \frac{P(n,k)}{k!}. \tag{3.2}$$

See page 210 for a table of values. Other formulas involving binomial coefficients include:

(a) $\binom{n}{k} = \dfrac{n(n-1)\cdots(n-k+1)}{k!} = \binom{n}{n-k}$

(b) $\binom{n}{0} = \binom{n}{n} = 1$ and $\binom{n}{1} = n$

(c) $\binom{n}{k} = \binom{n-1}{k} + \binom{n-1}{k-1}$

(d) $\binom{n}{0} + \binom{n}{1} + \cdots + \binom{n}{n} = 2^n$

(e) $\binom{n}{0} - \binom{n}{1} + \cdots + (-1)^n \binom{n}{n} = 0$

Example 3.8: For the 5 element set $\{a,b,c,d,e\}$ find the number of subsets containing exactly 3 elements.

Solution:

(S1) There are $\binom{5}{3} = \dfrac{5!}{3!2!} = 10$ subsets containing exactly 3 elements.

3.2. COMBINATORIAL METHODS

(S2) The subsets are

(a,b,c) (a,b,d) (a,b,e) (a,c,d) (a,c,e)
(a,d,e) (b,c,d) (b,c,e) (b,d,e) (c,d,e)

3.2.6 Sample selection

There are 4 ways in which a sample of k elements can be obtained from a set of n distinguishable objects.

Order counts?	Repetitions allowed?	The sample is called a	Number of ways to choose the sample
No	No	k-combination	$C(n,k)$
Yes	No	k-permutation	$P(n,k)$
No	Yes	k-combination with replacement	$C^R(n,k)$
Yes	Yes	k-permutation with replacement	$P^R(n,k)$

where

$$\begin{aligned} C(n,k) &= \binom{n}{k} = \frac{n!}{k!\,(n-k)!} \\ P(n,k) &= (n)_k = n^{\underline{k}} = \frac{n!}{(n-k)!} \\ C^R(n,k) &= C(n+k-1,k) = \frac{(n+k-1)!}{k!(n-1)!} \\ P^R(n,k) &= n^k \end{aligned} \qquad (3.3)$$

Example 3.9: There are 4 ways in which to choose a 2 element sample from the set $\{a,b\}$:

combination	$C(2,2) = 1$	ab
permutation	$P(2,2) = 2$	ab and ba
combination with replacement	$C^R(2,2) = 3$	aa, ab, and bb
permutation with replacement	$P^R(2,2) = 4$	aa, ab, ba, and bb

3.2.7 Balls into cells

There are 8 different ways in which n balls can be placed into k cells.

Distinguish balls?	Distinguish cells?	Can cells be empty?	Number of ways to place n balls into k cells
Yes	Yes	Yes	k^n
Yes	Yes	No	$k!\left\{{n \atop k}\right\}$
No	Yes	Yes	$C(k+n-1,n) = \binom{k+n-1}{n}$
No	Yes	No	$C(n-1,k-1) = \binom{n-1}{k-1}$
Yes	No	Yes	$\left\{{n \atop 1}\right\} + \left\{{n \atop 2}\right\} + \cdots + \left\{{n \atop k}\right\}$
Yes	No	No	$\left\{{n \atop k}\right\}$
No	No	Yes	$p_1(n) + p_2(n) + \cdots + p_k(n)$
No	No	No	$p_k(n)$

where $\left\{{n \atop k}\right\}$ is the Stirling cycle number and $p_k(n)$ is the number of partitions of the number n into exactly k integer pieces.

Given n distinguishable balls and k distinguishable cells, the number of ways in which we can place n_1 balls into cell 1, n_2 balls into cell 2, ..., n_k balls into cell k, is given by the multinomial coefficient $\binom{n}{n_1,n_2,\ldots,n_k}$.

3.2.8 Multinomial coefficients

The multinomial coefficient, $\binom{n}{n_1,n_2,\ldots,n_k} = C(n; n_1, n_2, \ldots, n_k)$, is the number of ways of choosing n_1 objects, then n_2 objects, ..., then n_k objects from a collection of n distinct objects without regard to order. This requires that $\sum_{j=1}^{k} n_j = n$.

Other ways to interpret the multinomial coefficient:

(1) Permutations (all objects not distinct): Given n_1 objects of one kind, n_2 objects of a second kind, ..., n_k objects of a kth kind, and $n_1 + n_2 + \cdots + n_k = n$. The number of permutations of the n objects is $\binom{n}{n_1,n_2,\ldots,n_k}$.

(2) Partitions: The number of ways of partitioning a set of n distinct objects into k subsets with n_1 objects in the first subset, n_2 objects in the second subset, ..., and n_k objects in the kth subset is $\binom{n}{n_1,n_2,\ldots,n_k}$.

The multinomial symbol is numerically evaluated as

$$\binom{n}{n_1, n_2, \ldots, n_k} = \frac{n!}{n_1! n_2! \cdots n_k!} \qquad (3.4)$$

Example 3.10: The number of ways to choose 2 objects, then 1 object, then 1 object from the set $\{a,b,c,d\}$ is $\binom{4}{2,1,1} = 12$; they are as follows (commas separate the ordered selections):

$$\begin{array}{llll}
\{ab,c,d\} & \{ab,d,c\} & \{ac,b,d\} & \{ac,d,b\} \\
\{ad,b,c\} & \{ad,c,b\} & \{bc,a,d\} & \{bc,d,a\} \\
\{bd,a,c\} & \{bd,c,a\} & \{cd,a,b\} & \{cd,b,a\}
\end{array}$$

3.2.9 Arrangements and derangements

(a) The number of ways to arrange n distinct objects in a row is $n!$; this is the number of permutations of n objects.

Example 3.11: For the three objects $\{a,b,c\}$ the number of arrangements is $3! = 6$. These permutations are $\{abc, bac, cab, acb, bca, cba\}$.

(b) The number of ways to arrange n non-distinct objects (assuming that there are k types of objects, and n_i copies of each object of type i) is the multinomial coefficient $\binom{n}{n_1, n_2, \ldots, n_k}$.

Example 3.12: For the set $\{a, a, b, c\}$ the parameters are $n = 4$, $k = 3$, $n_1 = 2$, $n_2 = 1$, and $n_3 = 1$. Hence, there are $\binom{4}{2,1,1} = \frac{4!}{2!\,1!\,1!} = 12$ arrangements, they are:

$$\begin{array}{cccccc} aabc & aacb & abac & abca & acab & acba \\ baac & baca & bcaa & caab & caba & cbaa \end{array}$$

(c) A *derangement* is a permutation of objects, in which object i is not in the ith location.

Example 3.13: All the derangements of $\{1, 2, 3, 4\}$ are:

$$\begin{array}{ccc} 2143 & 2341 & 2413 \\ 3142 & 3412 & 3421 \\ 4123 & 4312 & 4321 \end{array}$$

The number of derangements of n elements, D_n, satisfies the recursion relation: $D_n = (n-1)(D_{n-1} + D_{n-2})$, with the initial values $D_1 = 0$ and $D_2 = 1$. Hence,

$$D_n = n!\left(1 - \frac{1}{1!} + \frac{1}{2!} - \frac{1}{3!} + \cdots + (-1)^n \frac{1}{n!}\right)$$

The numbers D_n are also called *subfactorials* and *rencontres numbers*. For large values of n, $D_n/n! \sim e^{-1} \approx 0.37$. Hence, more than one of every three permutations is a derangement.

n	1	2	3	4	5	6	7	8	9	10
D_n	0	1	2	9	44	265	1854	14833	133496	1334961

3.3 PROBABILITY

The *sample space* of an experiment, denoted S, is the set of all possible outcomes. Each outcome of the sample space is also called an element of the sample space or a sample point. An *event* is any collection of outcomes contained in the sample space. A *simple event* consists of exactly one outcome and a *compound event* consists of more than one outcome.

3.3.1 Relative frequency concept of probability

Suppose an experiment is conducted n identical and independent times and $n(A)$ is the number of times the event A occurs. The quotient $n(A)/n$ is the relative frequency of occurrence of the event A. As n increases, the relative frequency converges to the limiting relative frequency of the event A. The probability of the event A, Prob $[A]$, is this limiting relative frequency.

3.3.2 Axioms of probability (discrete sample space)

(1) For any event A, Prob $[A] \geq 0$.

(2) Prob $[S] = 1$.

(3) If $A_1, A_2, A_3, \ldots$, is a finite or infinite collection of pairwise mutually exclusive events of S, then

$$\text{Prob}\,[A_1 \cup A_2 \cup A_3 \cup \cdots] = \text{Prob}\,[A_1] + \text{Prob}\,[A_2] + \text{Prob}\,[A_3] + \cdots \tag{3.5}$$

3.3.3 The probability of an event

The probability of an event A is the sum of Prob $[a_i]$ for all sample points a_i in the event A:

$$\text{Prob}\,[A] = \sum_{a_i \in A} \text{Prob}\,[a_i]. \tag{3.6}$$

If all of the outcomes in S are *equally likely*:

$$\text{Prob}\,[A] = \frac{n(A)}{n(S)} = \frac{\text{number of outcomes in } A}{\text{number of outcomes in } S}. \tag{3.7}$$

3.3.4 Probability theorems

(1) Prob $[\phi] = 0$ for any sample space S.

(2) If A and A' are complementary events, Prob $[A]$ + Prob $[A'] = 1$.

(3) For any events A and B, if $A \subset B$ then Prob $[A] \leq$ Prob $[B]$.

(4) For any events A and B,

$$\text{Prob}\,[A \cup B] = \text{Prob}\,[A] + \text{Prob}\,[B] - \text{Prob}\,[A \cap B]. \tag{3.8}$$

If A and B are mutually exclusive events, Prob $[A \cap B] = 0$ and

$$\text{Prob}\,[A \cup B] = \text{Prob}\,[A] + \text{Prob}\,[B]. \tag{3.9}$$

(5) For any events A and B,

$$\text{Prob}\,[A] = \text{Prob}\,[A \cap B] + \text{Prob}\,[A \cap B']. \tag{3.10}$$

(6) For any events A, B, and C,

$$\begin{aligned}\text{Prob}\,[A \cup B \cup C] =\;& \text{Prob}\,[A] + \text{Prob}\,[B] + \text{Prob}\,[C] \\ & - \text{Prob}\,[A \cap B] - \text{Prob}\,[A \cap C] - \text{Prob}\,[B \cap C] \\ & + \text{Prob}\,[A \cap B \cap C]. \end{aligned} \tag{3.11}$$

(7) For any events $A_1, A_2, \ldots, A_n$,

$$\text{Prob}\left[\bigcup_{i=1}^{n} A_i\right] \leq \sum_{i=1}^{n} \text{Prob}\,[A_i]. \tag{3.12}$$

3.3. PROBABILITY

Equality holds if the events are pairwise mutually exclusive.

3.3.5 Probability and odds

If the probability of an event A is $\text{Prob}[A]$ then

$$\text{odds for } A = \text{Prob}[A]/\text{Prob}[A'], \quad \text{Prob}[A'] \neq 0$$
$$\text{odds against } A = \text{Prob}[A']/\text{Prob}[A], \quad \text{Prob}[A] \neq 0. \quad (3.13)$$

If the odds for the event A are a:b, then $\text{Prob}[A] = a/(a+b)$.

Example 3.14: The odds of a fair coin coming up heads are 1:1; that it, is has a probability of ½.

The odds of a die showing a "1" are 5:1 against; that it, there is a probability of ⅚ that a "1" does not appear.

3.3.6 Conditional probability

The conditional probability of A given the event B has occurred is

$$\text{Prob}[A|B] = \frac{\text{Prob}[A \cap B]}{\text{Prob}[B]}, \quad \text{Prob}[B] > 0. \quad (3.14)$$

(1) If $\text{Prob}[A_1 \cap A_2 \cap \cdots \cap A_{n-1}] > 0$ then

$$\text{Prob}[A_1 \cap A_2 \cap \cdots \cap A_n] = \text{Prob}[A_1] \cdot \text{Prob}[A_2|A_1]$$
$$\cdot \text{Prob}[A_3|A_1 \cap A_2] \quad (3.15)$$
$$\cdots \text{Prob}[A_n|A_1 \cap A_2 \cap \cdots \cap A_{n-1}].$$

(2) If $A \subset B$, then $\text{Prob}[A|B] = \text{Prob}[A]/\text{Prob}[B]$ and $\text{Prob}[B|A] = 1$.

(3) $\text{Prob}[A'|B] = 1 - \text{Prob}[A|B]$.

Example 3.15: A local bank offers loans for three purposes: home (H), automobile (A), and personal (P), and two different types: fixed rate (FR) and adjustable rate (ADJ). The joint probability table given below presents the proportions for the various categories of loan and type:

		Loan Purpose		
		H	A	P
Type	FR	.27	.19	.14
	ADJ	.13	.09	.18

Suppose a person who took out a loan at this bank is selected at random.

(a) What is the probability the person has an automobile loan and it is fixed rate?

(b) Given the person has an adjustable rate loan, what is the probability it is for a home?

(c) Given the person does not have a personal loan, what is the probability it is adjustable rate?

Solution:

(S1) Prob$[A \cap FR] = .19$

(S2) Prob$[H \mid ADJ] = $ Prob$[H \cap ADJ]/$Prob$[ADJ] = .13/.4 = .325$

(S3) Prob$[ADJ \mid P'] = $ Prob$[ADJ \cap P']/$Prob$[P'] = .22/.68 = .3235$

3.3.7 The multiplication rule

$$\begin{aligned} \text{Prob}[A \cap B] &= \text{Prob}[A|B] \cdot \text{Prob}[B], \quad \text{Prob}[B] \neq 0 \\ &= \text{Prob}[B|A] \cdot \text{Prob}[A], \quad \text{Prob}[A] \neq 0 \end{aligned} \quad (3.16)$$

3.3.8 The law of total probability

Suppose $A_1, A_2, \ldots, A_n$ is a collection of mutually exclusive, exhaustive events, Prob$[A_i] \neq 0$, $i = 1, 2, \ldots, n$. For any event B:

$$\text{Prob}[B] = \sum_{i=1}^{n} \text{Prob}[B \mid A_i] \cdot \text{Prob}[A_i]. \quad (3.17)$$

3.3.9 Bayes' theorem

Suppose $A_1, A_2, \ldots, A_n$ is a collection of mutually exclusive, exhaustive events, Prob$[A_i] \neq 0$, $i = 1, 2, \ldots, n$. For any event B such that Prob$[B] \neq 0$:

$$\text{Prob}[A_k \mid B] = \frac{\text{Prob}[A_k \cap B]}{\text{Prob}[B]} = \frac{\text{Prob}[B \mid A_k] \cdot \text{Prob}[A_k]}{\sum_{i=1}^{n} \text{Prob}[B \mid A_i] \cdot \text{Prob}[A_i]}, \quad (3.18)$$

for $k = 1, 2, \ldots, n$.

Example 3.16: A large manufacturer uses three different trucking companies (A, B, and C) to deliver products. The probability a randomly selected shipment is delivered by each company is

$$\text{Prob}[A] = .60, \quad \text{Prob}[B] = .25, \quad \text{Prob}[C] = .15$$

Occasionally, a shipment is damaged (D) in transit.

$$\text{Prob}[D \mid A] = .01, \quad \text{Prob}[D \mid B] = .005, \quad \text{Prob}[D \mid C] = .015$$

Suppose a shipment is selected at random.

(a) Find the probability the shipment is sent by trucking company B and is damaged.

(b) Find the probability the shipment is damaged.

(c) Suppose a shipment arrives damaged. What is the probability it was shipped by company B?

Solution:

(S1) Prob$[B \cap D] = $ Prob$[B] \cdot $ Prob$[D \mid B] = (.25)(.005) = .00125$

(S2) $\text{Prob}[D] = \text{Prob}[A \cap D] + \text{Prob}[B \cap D] + \text{Prob}[C \cap D]$
$= \text{Prob}[A] \cdot \text{Prob}[D \mid A] + \text{Prob}[B] \cdot \text{Prob}[D \mid B]$
$+ \text{Prob}[B] \cdot \text{Prob}[D \mid B]$
$= (.60)(.01) + (.25)(.005) + (.15)(.015) = .0095$

(S3) $\text{Prob}[B \mid D] = \text{Prob}[B \cap D]/\text{Prob}[D] = .00125/.0095 = .1316$

3.3.10 Independence

(1) A and B are independent events if $\text{Prob}[A|B] = \text{Prob}[A]$ or, equivalently, if $\text{Prob}[B|A] = \text{Prob}[B]$.

(2) A and B are independent events if and only if $\text{Prob}[A \cap B] = \text{Prob}[A] \cdot \text{Prob}[B]$.

(3) $A_1, A_2, \ldots, A_n$ are pairwise independent events if

$$\text{Prob}[A_i \cap A_j] = \text{Prob}[A_i] \cdot \text{Prob}[A_j] \quad \text{for every pair } i, j \text{ with } i \neq j. \tag{3.19}$$

(4) $A_1, A_2, \ldots, A_n$ are mutually independent events if for every k, $k = 2, 3, \ldots, n$, and every subset of indices $i_1, i_2, \ldots, i_k$,

$$\text{Prob}[A_{i_1} \cap A_{i_2} \cap \cdots \cap A_{i_k}] = \text{Prob}[A_{i_1}] \cdot \text{Prob}[A_{i_2}] \cdots \text{Prob}[A_{i_k}]. \tag{3.20}$$

3.4 RANDOM VARIABLES

Given a sample space S, a random variable is a function with domain S and range some subset of the real numbers. A random variable is *discrete* if it can assume only a finite or countably infinite number of values. A random variable is *continuous* if its set of possible values is an entire interval of numbers. Random variables are denoted by upper-case letters, for example X.

3.4.1 Discrete random variables

3.4.1.1 Probability mass function

The probability distribution or probability mass function (pmf), $p(x)$, of a discrete random variable is a rule defined for every number x by $p(x) = \text{Prob}[X = x]$ such that

(1) $p(x) \geq 0$; and

(2) $\sum_{x} p(x) = 1$

3.4.1.2 Cumulative distribution function

The cumulative distribution function (cdf), $F(x)$, for a discrete random variable X with pmf $p(x)$ is defined for every number x:

$$F(x) = \text{Prob}[X \leq x] = \sum_{y \mid y \leq x} p(y). \tag{3.21}$$

(1) $\lim_{x \to -\infty} F(x) = 0$

(2) $\lim_{x \to \infty} F(x) = 1$

(3) If a and b are real numbers such that $a < b$, then $F(a) \leq F(b)$.

(4) $\text{Prob}\,[a \leq X \leq b] = \text{Prob}\,[X \leq b] - \text{Prob}\,[X < a] = F(b) - F(a^-)$ where a^- is the first value X assumes less than a. Valid for $a, b, \in \mathcal{R}$ and $a < b$.

3.4.2 Continuous random variables

3.4.2.1 Probability density function

The probability distribution or probability density function (pdf) of a continuous random variable X is a real-valued function $f(x)$ such that

$$\text{Prob}\,[a \leq X \leq b] = \int_a^b f(x)\,dx, \quad a, b \in \mathcal{R}, \quad a \leq b. \qquad (3.22)$$

(1) $f(x) \geq 0$ for $-\infty < x < \infty$

(2) $\int_{-\infty}^{\infty} f(x)\,dx = 1$

(3) $\text{Prob}\,[X = c] = 0$ for $c \in \mathcal{R}$.

3.4.2.2 Cumulative distribution function

The cumulative distribution function (cdf), $F(x)$, for a continuous random variable X is defined by

$$F(x) = \text{Prob}\,[X \leq x] = \int_{-\infty}^{x} f(y)\,dy \quad -\infty < x < \infty. \qquad (3.23)$$

(1) $\lim_{x \to -\infty} F(x) = 0$

(2) $\lim_{x \to \infty} F(x) = 1$

(3) If a and b are real numbers such that $a < b$, then $F(a) \leq F(b)$.

(4) $\text{Prob}\,[a \leq X \leq b] = \text{Prob}\,[X \leq b] - \text{Prob}\,[X < a] = F(b) - F(a)$, $a, b, \in \mathcal{R}$ and $a < b$.

(5) The pdf $f(x)$ may be found from the cdf:

$$f(x) = \frac{dF(x)}{dx} \quad \text{whenever the derivative exists.} \qquad (3.24)$$

3.4.3 Random functions

A random function of a real variable t is a function, denoted $X(t)$, that is a random variable for each value of t. If the variable t can assume any value in an interval, then $X(t)$ is called a *stochastic process*; if the variable t can only assume discrete values then $X(t)$ is called a *random sequence*.

3.5 MATHEMATICAL EXPECTATION

3.5.1 Expected value

(1) If X is a discrete random variable with pmf $p(x)$:

    (a) The expected value of X is
$$\mathrm{E}[X] = \mu = \sum_{x} x p(x), \qquad (3.25)$$

    (b) The expected value of a function $g(X)$ is
$$\mathrm{E}[g(X)] = \mu_{g(X)} = \sum_{x} g(x) p(x). \qquad (3.26)$$

(2) If X is a continuous random variable with pdf $f(x)$:

    (a) The expected value of X is
$$\mathrm{E}[X] = \mu = \int_{-\infty}^{\infty} x f(x)\, dx, \qquad (3.27)$$

    (b) The expected value of a function $g(X)$ is
$$\mathrm{E}[g(X)] = \mu_{g(X)} = \int_{-\infty}^{\infty} g(x) f(x)\, dx. \qquad (3.28)$$

(3) Jensen's inequality

Let $h(x)$ be a function such that $\dfrac{d^2}{dx^2}[h(x)] \geq 0$, then $\mathrm{E}[h(X)] \geq h(\mathrm{E}[X])$.

(4) Theorems:

    (a) $\mathrm{E}[aX + bY] = a\mathrm{E}[X] + b\mathrm{E}[Y]$

    (b) $\mathrm{E}[X \cdot Y] = \mathrm{E}[X] \cdot \mathrm{E}[Y]$ if X and Y are independent.

3.5.2 Variance

The variance of a random variable X is

$$\sigma^2 = \mathrm{E}\left[(X-\mu)^2\right] = \begin{cases} \displaystyle\sum_{x}(x-\mu)^2 p(x) & \text{if } X \text{ is discrete} \\ \displaystyle\int_{-\infty}^{\infty}(x-\mu)^2 f(x)\, dx & \text{if } X \text{ is continuous} \end{cases} \qquad (3.29)$$

The standard deviation of X is $\sigma = \sqrt{\sigma^2}$.

3.5.2.1 Theorems

Suppose X is a random variable, and a, b are constants.

(1) $\sigma_X^2 = \mathrm{E}[X^2] - (\mathrm{E}[X])^2$.

(2) $\sigma_{aX}^2 = a^2 \cdot \sigma_X^2, \qquad \sigma_{aX} = |a| \cdot \sigma_X$.

(3) $\sigma^2_{X+b} = \sigma^2_X$.

(4) $\sigma^2_{aX+b} = a^2 \cdot \sigma^2_X$, $\quad \sigma_{aX+b} = |a| \cdot \sigma_X$.

3.5.3 Moments

3.5.3.1 Moments about the origin

The moments about the origin completely characterize a probability distribution. The rth moment about the origin, $r = 0, 1, 2, \ldots$, of a random variable X is $\mu'_r = \mathrm{E}[X^r]$. The first moment about the origin is the mean of the random variable: $\mu'_1 = \mathrm{E}[X] = \mu$.

3.5.3.2 Moments about the mean

The rth moment about the mean, $r = 0, 1, 2, \ldots$, of a random variable X is $\mu_r = \mathrm{E}[(X - \mu)^r]$. The second moment about the mean is the variance of the random variable: $\mu_2 = \mathrm{E}[(X - \mu)^r] = \sigma^2 = \mu'_2 - \mu^2$.

3.5.3.3 Factorial moments

The rth factorial moment, $r = 0, 1, 2, \ldots$, of a random variable is $\mu_{[r]} = \mathrm{E}[X^{[r]}]$. where $x^{[r]}$ is the factorial expression

$$x^{[r]} = x(x-1)(x-2)\cdots(x-r+1). \tag{3.30}$$

3.5.4 Generating functions

3.5.4.1 Moment generating function

The moment generating function $m_X(t)$ is the expected value of e^{tX} and may be written as

$$\begin{aligned} m_X(t) &= \mathrm{E}\left[e^{tX}\right] \\ &= \mathrm{E}\left[1 + Xt + \frac{(Xt)^2}{2!} + \frac{(Xt)^3}{3!} + \cdots\right] \\ &= 1 + \mu'_1 t + \mu'_2 \frac{t^2}{2!} + \mu'_3 \frac{t^3}{3!} + \cdots \end{aligned} \tag{3.31}$$

The moments μ'_r are the coefficients of $t^r/r!$ in equation (3.31). Therefore, $m_X(t)$ *generates* the moments since the rth derivative of $m_X(t)$ evaluated at $t = 0$ yields μ'_r:

$$\mu'_r = m_x^{(r)}(0) = \left.\frac{d^r m_X(t)}{dt^r}\right|_{t=0} \tag{3.32}$$

Theorems: Suppose $m_X(t)$ is the moment generating function for the random variable X and a, b are constants.

(1) $m_{aX}(t) = m_X(at)$

(2) $m_{X+b}(t) = e^{bt} \cdot m_X(t)$

3.5. MATHEMATICAL EXPECTATION

(3) $m_{(X+b)/a}(t) = e^{(b/a)t} \cdot m_X(t/a)$

(4) If $X_1, X_2, \ldots, X_n$ are independent random variables and $Y = X_1 + X_2 + \cdots + X_n$, then $m_Y(t) = [m_X(t)]^n$.

The moment generating function for $X - \mu$ is

$$m_{X-\mu}(t) = e^{-\mu t} \cdot m_X(t). \tag{3.33}$$

Equation (3.33) may be used to generate the moments about the mean for the random variable X:

$$\mu_r = m_{X-\mu}^{(r)}(0) = \left. \frac{d^r \left(e^{-\mu t} \cdot m_X(t)\right)}{dt^r} \right|_{t=0} \tag{3.34}$$

3.5.4.2 Factorial moment generating functions

The factorial moment generating function of a random variable X is $P(t) = E\left[t^X\right]$. The rth derivative of the function $P\ t$, evaluated at $t = 1$ is the rth factorial moment. Therefore, the function P *generates* the factorial moments:

$$\mu_{[r]} = P^{(r)}(1) = \left. \frac{d^r P(t)}{dt^r} \right|_{t=1}. \tag{3.35}$$

In particular:

$$\begin{aligned} 1 &= P(1) \quad \text{"conservation of probability"} \\ \mu &= P'(1) \\ \sigma^2 &= P''(1) + P'(1) - [P'(1)]^2 \end{aligned} \tag{3.36}$$

3.5.4.3 Factorial moment generating function theorems

Theorems: Suppose $P_X(t)$ is the factorial moment generating function for the random variable X and a, b are constants.

(1) $P_{aX}(t) = P_X(t^a)$

(2) $P_{X+b}(t) = t^b \cdot P_X(t)$

(3) $P_{(X+b)/a}(t) = t^{b/a} \cdot P_X(t^{1/a})$

(4) $P_X(t) = m_X(\ln t)$, where $m_x(t)$ is the moment generating function for X.

(5) If $X_1, X_2, \ldots, X_n$ are *independent* random variables with factorial moment generating function $P_X(t)$ and $Y = X_1 + X_2 + \cdots + X_n$, then $P_Y(t) = [P_X(t)]^n$.

3.5.4.4 Cumulant generating function

Let $m_X(t)$ be a moment generating function. If $\ln m_X(t)$ can be expanded in the form

$$c(t) = \ln m_X(t) = \kappa_1 t + \kappa_2 \frac{t^2}{2!} + \kappa_3 \frac{t^3}{3!} + \cdots + \kappa_r \frac{t^r}{r!} + \cdots, \qquad (3.37)$$

then $c(t)$ is the cumulant generating function (or semi-invariant generating function). The constants κ_r are the cumulants (or semi-invariants) of the distribution. The rth derivative of c with respect to t, evaluated at 0 is the rth cumulant. The function c *generates* the cumulants:

$$\kappa_r = c^{(r)}(0) = \left. \frac{d^r c(t)}{dt^r} \right|_{t=0}. \qquad (3.38)$$

Marcienkiewicz's theorem states that either all but the first two cumulants vanish (i.e., it is a normal distribution) or there are an infinite number of non-vanishing cumulants.

3.5.4.5 Characteristic function

The characteristic function exists for every random variable X and is defined by $\phi(t) = \mathrm{E}\left[e^{itX}\right]$, where t is a real number and $i^2 = -1$. The rth derivative of ϕ with respect to t, evaluated at $t=0$ is $i^r \mu'_r$. Therefore, the characteristic function also generates the moments:

$$i^r \mu'_r = \phi^{(r)}(0) = \left. \frac{d^r \phi(t)}{dt^r} \right|_{t=0}. \qquad (3.39)$$

3.6 MULTIVARIATE DISTRIBUTIONS

3.6.1 Discrete case

A n-dimensional random variable $(X_1, X_2, \ldots, X_n)$ is n-dimensional discrete if it can assume only a finite or countably infinite number of values. The joint probability distribution, joint probability mass function, or joint density, for $(X_1, X_2, \ldots, X_n)$ is

$$p(x_1, x_2, \ldots, x_n) = \mathrm{Prob}\left[X_1 = x_1, X_2 = x_2, \ldots, X_n = x_n\right]$$
$$\forall (x_1, x_2, \ldots, x_n). \qquad (3.40)$$

Suppose E is a subset of values the random variable may assume. The probability the event E occurs is

$$\mathrm{Prob}\left[E\right] = \mathrm{Prob}\left[(X_1, X_2, \ldots, X_n) \in E\right]$$
$$= \sum \sum \cdots \sum_{(x_1, x_2, \ldots, x_n) \in E} p(x_1, x_2, \ldots, x_n). \qquad (3.41)$$

3.6. MULTIVARIATE DISTRIBUTIONS

The cumulative distribution function for $(X_1, X_2, \ldots, X_n)$ is

$$F(x_1, x_2, \ldots, x_n) = \sum_{t_1 | t_1 \leq x_1} \sum_{t_2 | t_2 \leq x_2} \cdots \sum_{t_n | t_n \leq x_n} p(x_1, x_2, \ldots, x_n). \quad (3.42)$$

3.6.2 Continuous case

The continuous random variables $X_1, X_2, \ldots, X_n$ are jointly distributed if there exists a function f such that $f(x_1, x_2, \ldots, x_n) \geq 0$ for $-\infty < x_i < \infty$, $i = 1, 2, \ldots, n$, and for any event E

$$\begin{aligned} \text{Prob}\,[E] &= \text{Prob}\,[(X_1, X_2, \ldots, X_n) \in E] \\ &= \iint \cdots \int_E f(x_1, x_2, \ldots, x_n)\, dx_n \cdots dx_1 \end{aligned} \quad (3.43)$$

where f is the joint distribution function or joint probability density function for the random variables $X_1, X_2, \ldots, X_n$. The cumulative distribution function for $X_1, X_2, \ldots, X_n$ is

$$F(x_1, x_2, \ldots, x_n) = \int_{-\infty}^{x_1} \int_{-\infty}^{x_2} \cdots \int_{-\infty}^{x_n} f(x_1, x_2, \ldots, x_n)\, dx_n \cdots dx_1. \quad (3.44)$$

Given the cumulative distribution function, F, the probability density function may be found by

$$f(x_1, x_2, \ldots, x_n) = \frac{\partial^n}{\partial x_1 \, \partial x_2 \cdots \partial x_n} F(x_1, x_2, \ldots, x_n) \quad (3.45)$$

wherever the partials exist.

3.6.3 Expectation

Let $g(X_1, X_2, \ldots, X_n)$ be a function of the random variables $X_1, \ldots, X_n$. The expected value of $g(X_1, X_2, \ldots, X_n)$ is

$$\text{E}\,[g(X_1, X_2, \ldots, X_n)] = \sum_{x_1} \sum_{x_2} \cdots \sum_{x_n} g(x_1, x_2, \ldots, x_n) p(x_1, x_2, \ldots, x_n) \quad (3.46)$$

if $X_1, X_2, \ldots, X_n$ are discrete, and

$$\begin{aligned} \text{E}\,[g(X_1, X_2, \ldots, X_n)] = \\ \int_{-\infty}^{\infty} \int_{-\infty}^{\infty} \cdots \int_{-\infty}^{\infty} g(x_1, \ldots, x_n) f(x_1, \ldots, x_n)\, dx_n \cdots dx_1 \end{aligned} \quad (3.47)$$

if $X_1, X_2, \ldots, X_n$ are continuous.

If $c_1, c_2, \ldots, c_n$ are constants, then

$$\text{E}\left[\sum_{i=1}^{n} c_i g_i(X_1, X_2, \ldots, X_n)\right] = \sum_{i=1}^{n} c_i \text{E}\,[g_i(X_1, X_2, \ldots, X_n)]. \quad (3.48)$$

3.6.4 Moments

If $X_1, X_2, \ldots, X_n$ are jointly distributed, the rth moment of X_i is $\mathrm{E}[X_i^r]$.

The joint (product) moments about the origin are $\mathrm{E}[X_1^{r_1} X_2^{r_2} \cdots X_n^{r_n}]$ if $X_1, X_2, \ldots, X_n$ are continuous. The value $r = r_1 + r_2 + \cdots + r_n$ is the *order* of the moment.

If $\mathrm{E}[X_i] = \mu_i$, then the joint moments about the mean are
$\mathrm{E}[(X_1 - \mu_1)^{r_1} (X_2 - \mu_2)^{r_2} \cdots (X_n - \mu_n)^{r_n}]$.

3.6.5 Marginal distributions

Let $X_1, X_2, \ldots, X_n$ be a collection of random variables. The marginal distribution of a subset of the random variables $X_1, X_2, \ldots, X_k$ (with $(k < n)$) is

$$g(x_1, x_2, \ldots, x_k) = \sum_{x_{k+1}} \sum_{x_{k+2}} \cdots \sum_{x_n} p(x_1, x_2, \ldots, x_n) \qquad (3.49)$$

if $X_1, X_2, \ldots, X_n$ are discrete, and

$$g(x_1, x_2, \ldots, x_k) = \int_{-\infty}^{\infty} \int_{-\infty}^{\infty} \cdots \int_{-\infty}^{\infty} f(x_1, x_2, \ldots, x_n) \, dx_{k+1} dx_{k+2} \cdots dx_n \qquad (3.50)$$

if $X_1, X_2, \ldots, X_n$ are continuous.

Example 3.17: The joint density functions $g(x,y) = x+y$ and $h(x,y) = (x+\frac{1}{2})(y+\frac{1}{2})$ when $0 \leq x \leq 1$ and $0 \leq y \leq 1$ have the same marginal distributions. Using equation (3.50):

$$\begin{aligned} g_x(x) &= \int_0^1 g(x,y) \, dy = \left(xy + \frac{y^2}{2} \right) \bigg|_{y=0}^{y=1} = x + \frac{1}{2} \\ h_x(x) &= \int_0^1 h(x,y) \, dy = \left(x + \frac{1}{2} \right) \left(\frac{y^2}{2} + \frac{y}{2} \right) \bigg|_{y=0}^{y=1} = x + \frac{1}{2} \end{aligned} \qquad (3.51)$$

and, by symmetry, $g_y(y)$ has the same form as $g_x(x)$ (likewise for $h_y(y)$ and $h_x(x)$).

3.6.6 Independent random variables

Let $X_1, X_2, \ldots, X_n$ be a collection of discrete random variables with joint probability distribution function $p(x_1, x_2, \ldots, x_n)$. Let $g_{X_i}(x_i)$ be the marginal distribution for X_i. The random variables $X_1, X_2, \ldots, X_n$ are independent if and only if

$$p(x_1, x_2, \ldots, x_n) = g_{X_1}(x_1) \cdot g_{X_2}(x_2) \cdots g_{X_n}(x_n). \qquad (3.52)$$

Let $X_1, X_2, \ldots, X_n$ be a collection of continuous random variables with joint probability distribution function $f(x_1, x_2, \ldots, x_n)$. Let $g_{X_i}(x_i)$ be the marginal distribution for X_i. The random variables $X_1, X_2, \ldots, X_n$ are independent if and only if

$$f(x_1, x_2, \ldots, x_n) = g_{X_1}(x_1) \cdot g_{X_2}(x_2) \cdots g_{X_n}(x_n). \qquad (3.53)$$

3.6. MULTIVARIATE DISTRIBUTIONS

Example 3.18: Suppose X_1, X_2, and X_3 are independent random variables with probability density functions given by

$$g_{X_1}(x_1) = \begin{cases} e^{-x_1} & x_1 > 0 \\ 0 & \text{elsewhere} \end{cases}$$

$$g_{X_2}(x_2) = \begin{cases} 3e^{-3x_2} & x_2 > 0 \\ 0 & \text{elsewhere} \end{cases} \qquad g_{X_3}(x_3) = \begin{cases} 7e^{-7x_3} & x_3 > 0 \\ 0 & \text{elsewhere} \end{cases}$$

Using equation (3.53), the joint probability distribution for X_1, X_2, X_3 is

$$\begin{aligned} f(x_1, x_2, x_3) &= g_{X_1}(x_1) \cdot g_{X_2}(x_2) \cdot g_{X_3}(x_3) \\ &= (e^{-x_1}) \cdot (3e^{-3x_2}) \cdot (7e^{-7x_3}) \\ &= 21 e^{-x_1 - 3x_2 - 7x_3} \qquad [x_1 > 0, x_2 > 0, x_3 > 0]. \end{aligned}$$

3.6.7 Conditional distributions

Let $X_1, X_2, \ldots, X_n$ be a collection of random variables. The conditional distribution of any subset of the random variables $X_1, X_2, \ldots, X_k$ given $X_{k+1} = x_{k+1}, X_{k+2} = x_{k+2}, \ldots, X_n = x_n$ is

$$p(x_1, x_2, \ldots, x_k \mid x_{k+1}, x_{k+2} \ldots, x_n) = \frac{p(x_1, x_2, \ldots, x_n)}{g(x_{k+1}, x_{k+2}, \ldots, x_n)} \qquad (3.54)$$

if $X_1, X_2, \ldots, X_n$ are discrete with joint distribution function $p(x_1, x_2, \ldots, x_n)$ and $X_{k+1}, X_{k+2}, \ldots, X_n$ have marginal distribution $g(x_{k+1}, x_{k+2}, \ldots, x_n) \neq 0$, and

$$f(x_1, x_2, \ldots, x_k \mid x_{k+1}, x_{k+2}, \ldots, x_n) = \frac{f(x_1, x_2, \ldots, x_n)}{g(x_{k+1}, x_{k+2}, \ldots, x_n)} \qquad (3.55)$$

if $X_1, X_2, \ldots, X_n$ are continuous with joint distribution function $f(x_1, x_2, \ldots, x_n)$ and $X_{k+1}, X_{k+2}, \ldots, X_n$ have marginal distribution $g(x_{k+1}, x_{k+2}, \ldots, x_n) \neq 0$.

Example 3.19: Suppose X_1, X_2, X_3 have a joint distribution function given by

$$f(x_1, x_2, x_3) = \begin{cases} (x_1 + x_2)e^{-x_3} & \text{when } 0 < x_1 < 1,\ 0 < x_2 < 1,\ x_3 > 0 \\ 0 & \text{elsewhere} \end{cases}$$

The marginal distribution of X_2 is

$$\begin{aligned} g(x_2) &= \int_0^1 \int_0^\infty (x_1 + x_2) e^{-x_3} \, dx_3 \, dx_1 \\ &= \int_0^1 (x_1 + x_2) \, dx_1 = \frac{1}{2} + x_2, \qquad 0 < x_2 < 1. \end{aligned}$$

The conditional distribution of X_1, X_3 given $X_2 = x_2$ is

$$f(x_1, x_3 \mid x_2) = \frac{f(x_1, x_2, x_3)}{g(x_2)} = \frac{(x_1 + x_2)e^{-x_3}}{\frac{1}{2} + x_2}.$$

If $X_2 = 3/4$, then

$$f(x_1, x_3 \mid 3/4) = \frac{\left(x_1 + \frac{3}{4}\right) e^{-x_3}}{\frac{1}{2} + \frac{3}{4}} = \frac{4}{5} \left(x_1 + \frac{3}{4}\right) e^{-x_3}, \quad 0 < x_1 < 1, \; x_3 > 0.$$

3.6.8 Variance and covariance

Let $X_1, X_2, \ldots, X_n$ be a collection of random variables. The variance, σ_{ii}, of X_i is

$$\sigma_{ii} = \sigma_i^2 = \mathrm{E}\left[(X_i - \mu_i)^2\right] \tag{3.56}$$

and the covariance, σ_{ij}, of X_i and X_j is

$$\sigma_{ij} = \rho_{ij} \sigma_i \sigma_j = \mathrm{E}\left[(X_i - \mu_i)(X_j - \mu_j)\right] \tag{3.57}$$

where ρ_{ij} is the correlation coefficient and σ_i and σ_j are the standard deviations of X_i and X_j, respectively.

Theorems:

(1) If $X_1, X_2, \ldots, X_n$ are independent, then

$$\mathrm{E}[X_1 X_2 \cdots X_n] = \mathrm{E}[X_1]\mathrm{E}[X_2] \cdots \mathrm{E}[X_n]. \tag{3.58}$$

(2) For two random variables X_i and X_j:

$$\sigma_{ij} = \mathrm{E}[X_i X_j] - \mathrm{E}[X_i]\mathrm{E}[X_j]. \tag{3.59}$$

(3) If X_i and X_j are independent random variables, then $\sigma_{ij} = 0$.

(4) Two variables may be dependent and have zero covariance. For example, let X take the four values $\{-2, -1, 1, 2\}$ with equal probability. If $Y = X^2$ then the covariance of X and Y is zero.

3.6.9 Correlation coefficient

The correlation coefficient, defined by (see equation (3.57))

$$\rho_{ij} = \frac{\sigma_{ij}}{\sigma_i \sigma_j} \tag{3.60}$$

is no greater than one in magnitude: $|\rho_{ij}| \leq 1$. Figure 3.5 contains 4 data sets of 100 points each; the correlation coefficients vary from -0.7 to 0.99.

3.6.10 Moment generating function

Let $X_1, X_2, \ldots, X_n$ be a collection of random variables. The joint moment generating function is

$$m(t_1, t_2, \ldots, t_n) = m(\mathbf{t}) = \mathrm{E}\left[e^{t_1 X_1 + t_2 X_2 + \cdots + t_n X_n}\right] = \mathrm{E}\left[e^{\mathbf{t} \cdot \mathbf{X}}\right] \tag{3.61}$$

if it exists for all values of t_i such that $|t_i| < h^2$ (for some value h).

The rth moment of X_i may be obtained (generated) by differentiating $m(t_1, t_2, \ldots, t_n)$ r times with respect to t_i, and then evaluating the result with

3.6. MULTIVARIATE DISTRIBUTIONS

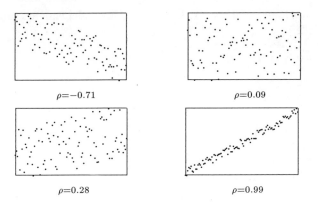

Figure 3.5: Data sets illustrating different correlation coefficients.

all t's equal to zero:

$$\mathrm{E}\left[X_i^r\right] = \left.\frac{\partial^r m(t_1, t_2, \ldots, t_n)}{\partial t_i^r}\right|_{(t_1, t_2, \ldots, t_n) = (0, 0, \ldots, 0)} \quad (3.62)$$

The rth joint moment, $r = r_1 + r_2 + \cdots + r_n$, may be obtained by differentiating $m(t_1, t_2, \ldots, t_n)$ r_1 times with respect to t_1, r_2 times with respect to t_2, $\ldots$, and r_n times with respect to t_n, and then evaluating the result with all t's equal to zero:

$$\mathrm{E}\left[X_1^{r_1} X_2^{r_2} \cdots X_n^{r_n}\right] = \left.\frac{\partial^r m(t_1, t_2, \ldots, t_n)}{\partial t_1^{r_1} \partial t_2^{r_2} \cdots \partial t_n^{r_n}}\right|_{(t_1, t_2, \ldots, t_n) = (0, 0, \ldots, 0)} \quad (3.63)$$

3.6.11 Linear combination of random variables

Let $X_1, X_2, \ldots, X_m$ and $Y_1, Y_2, \ldots, Y_n$ be random variables, let $a_1, a_2, \ldots, a_m$ and $b_1, b_2, \ldots, b_n$ be constants, and let U and V be the linear combinations

$$U = \sum_{i=1}^{m} a_i X_i, \qquad V = \sum_{j=1}^{n} b_j Y_j. \quad (3.64)$$

Theorems:

(1) $\mathrm{E}[U] = \sum_{i=1}^{m} a_i \mathrm{E}[X_i]$.

(2) $\sigma_i^2 = \sum_{i=1}^{m} a_i^2 \sigma_i^2 + 2 \sum\sum_{i<j} a_i a_j \sigma_{ij}$,

where the double sum extends over all pairs (i, j) with $i < j$.

(3) If the random variables $X_1, X_2, \ldots, X_m$ are independent, then
$$\sigma_U^2 = \sum_{i=1}^m a_i^2 \sigma_i^2.$$

(4) $\sigma_{UV} = \displaystyle\sum_{i=1}^m \sum_{j=1}^n a_i b_j \sigma_{ij}.$

3.7 INEQUALITIES

1. *Bienaymé–Chebyshev inequality*: If $E[|X|^r] < \infty$ for all $r > 0$ (r not necessarily an integer) then, for every $a > 0$
$$\text{Prob}\,[|X| \geq a] \leq \frac{E[|X|^r]}{a^r} \tag{3.65}$$

2. *Cauchy–Schwartz inequality*: Let X and Y be random variables in which $E[Y^2]$ and $E[Z^2]$ exist, then
$$(E[YZ])^2 \leq E[Y^2]\,E[Z^2] \tag{3.66}$$

3. *Chebyshev inequality*: Let c be any real number and let X be a random variable for which $E[(X-c)^2]$ is finite. Then for every $\epsilon > 0$ the following holds
$$\text{Prob}\,[|X - c| \geq \epsilon] \leq \frac{1}{\epsilon^2} E\left[(X-c)^2\right] \tag{3.67}$$

4. *Chebyshev inequality (one-sided)*: Let X be a random variable with zero mean (i.e., $E[X] = 0$) and variance σ^2. Then for any positive a
$$\text{Prob}\,[X > a] \leq \frac{\sigma^2}{\sigma^2 + a^2} \tag{3.68}$$

5. *Jensen inequality*: If $E[X]$ exists, and if $f(x)$ is a convex $\cup$ ("convex cup") function, then
$$E[f(X)] \geq f(E[X]) \tag{3.69}$$

6. *Kolmogorov inequality*: Let $X_1, X_2, \ldots, X_n$ be n independent random variables such that $E[X_i] = 0$ and $\text{Var}(X_i) = \sigma_{X_i}^2$ is finite. Then, for all $a > 0$,
$$\text{Prob}\left[\max_{i=1,\ldots,n} |X_1 + X_2 + \cdots + X_i| > a\right] \leq \sum_{i=1}^n \frac{\sigma_i^2}{a^2} \tag{3.70}$$

7. *Markov inequality*: If X is random variable which takes only nonnegative values, then for any $a > 0$
$$\text{Prob}\,[X \geq a] \leq \frac{E[X]}{a} \tag{3.71}$$

CHAPTER 4

Functions of Random Variables

Let $X_1, X_2, \ldots, X_n$ be a collection of random variables with joint probability mass function $p(x_1, x_2, \ldots, x_n)$ (if the collection is discrete) or joint density function $f(x_1, x_2, \ldots, x_n)$ (if the collection is continuous). Suppose the random variable $Y = Y(X_1, X_2, \ldots, X_n)$ is a function of $X_1, X_2, \ldots, X_n$. Methods for finding the distribution of Y are presented below and sampling distributions are discussed in the following section.

4.1 FINDING THE PROBABILITY DISTRIBUTION

The following techniques may be used to determine the probability distribution for $Y = Y(X_1, X_2, \ldots, X_n)$.

4.1.1 Method of distribution functions

Let $X_1, X_2, \ldots, X_n$ be a collection of continuous random variables.

(1) Determine the region $Y = y$.
(2) Determine the region $Y \leq y$.
(3) Compute $F(y) = \text{Prob}\,[Y \leq y]$ by integrating the joint density function $f(x_1, x_2, \ldots, x_n)$ over the region $Y \leq y$.
(4) Compute the probability density function for Y, $f(y)$, by differentiating $F(y)$:

$$f(y) = \frac{dF(y)}{dy}. \tag{4.1}$$

Example 4.20: Suppose the joint density function of X_1 and X_2 is given by

$$f(x_1, x_2) = \begin{cases} 4x_1 x_2 e^{-(x_1^2 + x_2^2)} & \text{for } x_1 > 0,\ x_2 > 0 \\ 0 & \text{elsewhere} \end{cases}$$

and $Y = \sqrt{X_1^2 + X_2^2}$. Find the cumulative distribution function for Y and the probability density function for Y.

Solution:

(S1) The region $Y \leq y$ is a quarter circle in quadrant I, shown shaded in Figure 4.1.

(S2) The cumulative distribution function for Y is given by

$$F(y) = \int_0^y \int_0^{\sqrt{y^2-x_1^2}} 4x_1 x_2 e^{-(x_1^2+x_2^2)} \, dx_2 \, dx_1$$

$$= \int_0^y 2x_1 (e^{-x_1^2} - e^{-y^2}) \, dx_1 \qquad (4.2)$$

$$= 1 - (1+y^2)e^{-y^2}$$

(S3) The probability density function for Y is given by

$$f(y) = F'(y) = -[(2y)e^{-y^2} + (1+y^2)(-2y)e^{-y^2}]$$

$$= 2y^3 e^{-y^2}, \quad \text{when } y > 0 \qquad (4.3)$$

4.1.2 Method of transformations (one variable)

Let X be a continuous random variable with probability density function $f_X(x)$. If $u(x)$ is differentiable and either increasing or decreasing, then $Y = u(X)$ has probability density function

$$f_Y(y) = f_X(w(y)) \cdot |w'(y)|, \qquad u'(x) \neq 0 \qquad (4.4)$$

where $x = w(y) = u^{-1}(y)$.

Example 4.21: Let X be a standard normal random variable, and let $Y = X^2$. What is the distribution of Y?

Solution:

(S1) Since X can be both positive and negative, two regions of X correspond to the same value of Y.

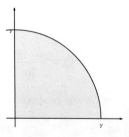

Figure 4.1: Integration region for example 4.20.

4.1. FINDING THE PROBABILITY DISTRIBUTION

(S2) The computation is

$$f_y(y) = \left[f_x(x) + f_x(-x)\right] \left|\frac{dy}{dx}\right|$$

$$= \left[\frac{e^{-x^2/2}}{\sqrt{2\pi}} + \frac{e^{-(-x)^2/2}}{\sqrt{2\pi}} + \right] \frac{1}{2\sqrt{y}} \quad (4.5)$$

$$= \frac{1}{\sqrt{2\pi y}} e^{-x^2/2}$$

$$= \frac{1}{\sqrt{2\pi y}} e^{-y/2}$$

which is the probability density function for a chi-square random variable with one degree of freedom.

Example 4.22: Given two independent random variables X and Y with joint probability density $f(x,y)$, let $U = X/Y$ be the ratio distribution. The probability density is:

$$f_U(u) = \int_{-\infty}^{\infty} |x| f(x, ux) \, dx \quad (4.6)$$

If X and Y are normally distributed, then U has a Cauchy distribution. If X and Y are uniformly distributed on $[0, 1]$, then

$$f_U(u) = \begin{cases} 0 & \text{for } u < 0 \\ 1/2 & \text{for } 0 \leq u \leq 1 \\ \frac{1}{2u^2} & \text{for } u > 1 \end{cases} \quad (4.7)$$

4.1.3 Method of transformations (two or more variables)

Let X_1 and X_2 be continuous random variables with joint density function $f(x_1, x_2)$. Let the functions $y_1 = u_1(x_1, x_2)$ and $y_2 = u_2(x_1, x_2)$ represent a one-to-one transformation from the x's to the y's and let the partial derivatives with respect to both x_1 and x_2 exist. The joint density function of $Y_1 = u_1(X_1, X_2)$ and $Y_2 = u_2(X_1, X_2)$ is

$$g(y_1, y_2) = f(w_1(y_1, y_2), w_2(y_1, y_2)) \cdot |J| \quad (4.8)$$

where $y_1 = u_1(x_1, x_2)$ and $y_2 = u_2(x_1, x_2)$ are uniquely solved for $x_1 = w_1(y_1, y_2)$ and $x_2 = w_2(y_1, y_2)$, and J is the determinant of the Jacobian

$$J = \begin{vmatrix} \frac{\partial x_1}{\partial y_1} & \frac{\partial x_1}{\partial y_2} \\ \frac{\partial x_2}{\partial y_1} & \frac{\partial x_2}{\partial y_2} \end{vmatrix}. \quad (4.9)$$

This method of transformations may be extended to functions of n random variables. Let $X_1, X_2, \ldots, X_n$ be continuous random variables with joint density function $f(x_1, x_2, \ldots, x_n)$. Let the functions $y_1 = u_1(x_1, x_2, \ldots, x_n)$, $y_2 = u_2(x_1, x_2, \ldots, x_n)$, ..., $y_n = u_n(x_1, x_2, \ldots, x_n)$ represent a one-to-one transformation from the x's to the y's and let the partial derivatives with respect to $x_1, x_2, \ldots, x_n$ exist. The joint density function of $Y_1 = u_1(X_1, X_2, \ldots, X_n)$, $Y_2 = u_2(X_1, X_2, \ldots, X_n), \ldots, Y_n = u_n(X_1, X_2, \ldots, X_n)$ is

$$g(y_1, y_2, \ldots, y_n) = f(w_1(y_1, \ldots, y_n), \ldots, w_n(y_1, \ldots, y_n)) \cdot |J| \qquad (4.10)$$

where the functions $y_1 = u_1(x_1, x_2, \ldots, x_n)$, $y_2 = u_2(x_1, x_2, \ldots, x_n), \ldots$, $y_n = u_n(x_1, x_2, \ldots, x_n)$ are uniquely solved for $x_1 = w_1(y_1, y_2, \ldots, y_n)$, $x_2 = w_2(y_1, y_2, \ldots, y_n), \ldots, x_n = w_n(y_1, y_2, \ldots, y_n)$ and J is the determinant of the Jacobian

$$J = \begin{vmatrix} \frac{\partial x_1}{\partial y_1} & \frac{\partial x_1}{\partial y_2} & \cdots & \frac{\partial x_1}{\partial y_n} \\ \frac{\partial x_2}{\partial y_1} & \frac{\partial x_2}{\partial y_2} & \cdots & \frac{\partial x_2}{\partial y_n} \\ \vdots & \vdots & \ddots & \vdots \\ \frac{\partial x_n}{\partial y_1} & \frac{\partial x_n}{\partial y_2} & \cdots & \frac{\partial x_n}{\partial y_n} \end{vmatrix} \qquad (4.11)$$

Example 4.23: Suppose the random variables X and Y are independent with probability density functions $f_X(x)$ and $f_Y(y)$, then the probability density of their sum, $Z = X + Y$, is given by

$$f_Z(z) = \int_{-\infty}^{\infty} f_X(t) f_Y(z - t) \, dt \qquad (4.12)$$

Example 4.24: Suppose the random variables X and Y are independent with probability density functions $f_X(x)$ and $f_Y(y)$, then the probability density of their product, $Z = XY$, is given by

$$f_Z(z) = \int_{-\infty}^{\infty} \frac{1}{|t|} f_X(t) f_Y\left(\frac{z}{t}\right) dt \qquad (4.13)$$

Example 4.25: Two random variables X and Y have a joint normal distribution. The probability density is $f(x, y) = \frac{1}{2\pi\sigma^2} \exp\left(\frac{x^2 + y^2}{2\sigma^2}\right)$. Find the probability density of the system (R, Φ) if

$$\begin{aligned} X &= R \cos \Phi \\ Y &= R \sin \Phi \end{aligned} \qquad (4.14)$$

Solution:

(S1) Use $f(r, \phi) = f\left[x(r, \phi), y(r, \phi)\right] \left|\frac{\partial(x,y)}{\partial(r,\phi)}\right|$ and $\left|\frac{\partial(x,y)}{\partial(r,\phi)}\right| = r$.

4.2. SUMS OF RANDOM VARIABLES

(S2) Then

$$f(r,\phi) = \frac{r}{2\pi\sigma^2} \exp\left(-\frac{r^2 \cos^2\phi + r^2 \sin^2\phi}{2\sigma^2}\right)$$

$$= \underbrace{\frac{1}{2\pi}}_{f_\Phi(\phi)} \underbrace{\frac{r}{\sigma^2} \exp\left(-\frac{r^2}{2\sigma^2}\right)}_{f_R(r)} \qquad (4.15)$$

where $f_R(r)$ is a Rayleigh distribution and $f_\Phi(\phi)$ is a uniform distribution.

4.1.4 Method of moment generating functions

To determine the distribution of Y:

(1) Determine the moment generating function for Y, $m_Y(t)$.

(2) Compare $m_Y(t)$ with known moment generating functions. If $m_Y(t) = m_U(t)$ for all t, then Y and U have identical distributions.

Theorems:

(1) Let X and Y be random variables with moment generating functions $m_X(t)$ and $m_Y(t)$, respectively. If $m_X(t) = m_Y(t)$ for all t, then X and Y have the same probability distributions.

(2) Let $X_1, X_2, \ldots, X_n$ be independent random variables and let $Y = X_1 + X_2 + \cdots + X_n$, then

$$m_Y(t) = \prod_{i=1}^{n} m_{X_i}(t). \qquad (4.16)$$

4.2 SUMS OF RANDOM VARIABLES

4.2.1 Deterministic sums of random variables

If $Y = X_1 + X_2 + \cdots + X_n$ and

(a) the $X_1, X_2, \ldots, X_n$ are independent random variables with factorial moment generating functions $P_{X_i}(t)$, then

$$P_Y(t) = \prod_{i=1}^{n} P_{X_i}(t) \qquad (4.17)$$

(b) the $X_1, X_2, \ldots, X_n$ are independent random variables with the same factorial moment generating function $P_X(t)$, then

$$P_Y(t) = [P_X(t)]^n \qquad (4.18)$$

(c) the $X_1, X_2, \ldots, X_n$ are independent random variables with characteristic functions $\phi_{X_i}(t)$, then

$$\phi_Y(t) = \prod_{i=1}^{n} \phi_{X_i}(t) \qquad (4.19)$$

(d) the $X_1, X_2, \ldots, X_n$ are independent random variables with the same characteristic function $\phi_X(t)$, then

$$\phi_Y(t) = [\phi_X(t)]^n \qquad (4.20)$$

Example 4.26: What is the distribution of the sum of two normal random variables?
Solution:
(S1) Let X_1 be $N(\mu_1, \sigma_1)$ and let X_2 be $N(\mu_2, \sigma_2)$.
(S2) The characteristic functions are (see page 99) $\phi_{X_1}(t) = \exp\left(\mu_1 it - \frac{\sigma_1^2 t}{2}\right)$ and $\phi_{X_2}(t) = \exp\left(\mu_2 it - \frac{\sigma_2^2 t}{2}\right)$.
(S3) From equation (4.19) the characteristic function for $Y = X_1 + X_2$ is

$$\phi_Y(t) = \phi_{X_1}(t) \cdot \phi_{X_2}(t) = \exp\left((\mu_1 + \mu_2)it - \frac{(\sigma_1^2 + \sigma_2^2)t}{2}\right) \qquad (4.21)$$

(S4) This last expression is the characteristic function for a normal random variable with mean $\mu_Y = \mu_1 + \mu_2$ and variance of $\sigma_Y^2 = \sigma_1^2 + \sigma_2^2$.
(S5) Conclusion: the distribution of the sum of two normal random variables is normal; the means add and the variances add.

See section 3.6.11 for linear combinations of random variables.

4.2.2 Random sums of random variables

If $T = \sum_{i=1}^{N} X_i$ where N is an integer valued random variable with factorial generating function $P_N(t)$, the $\{X_i\}$ are discrete independent and identically distributed random variables with factorial generating function $P_X(t)$, and the $\{X_i\}$ are independent of N, then the factorial generating function for T is

$$P_T(t) = P_N(P_X(t)) \qquad (4.22)$$

(If the $\{X_i\}$ are continuous random variables, then $\phi_T(t) = P_N(\phi_X(t))$.)
Hence (using equation (3.36))

$$\begin{aligned}\mu_T &= \mu_N \mu_X \\ \sigma_T^2 &= \mu_N \sigma_X^2 + \mu_X \sigma_N^2\end{aligned} \qquad (4.23)$$

4.3 SAMPLING DISTRIBUTIONS

4.3.1 Definitions

(1) The random variables $X_1, X_2, \ldots, X_n$ are a random sample of size n from an infinite population if $X_1, X_2, \ldots, X_n$ are independent and identically distributed (iid).

4.3. SAMPLING DISTRIBUTIONS

(2) If $X_1, X_2, \ldots, X_n$ are a random sample, then the sample total and sample mean are

$$T = \sum_{i=1}^{n} X_i \quad \text{and} \quad \overline{X} = \frac{1}{n} \sum_{i=1}^{n} X_i, \qquad (4.24)$$

respectively. The sample variance is

$$S^2 = \frac{1}{n-1} \sum_{i=1}^{n} (X_i - \overline{X})^2. \qquad (4.25)$$

4.3.2 The sample mean

Consider an infinite population with mean μ, variance σ^2, skewness γ_1, and kurtosis γ_2. Using a sample of size n, the parameters describing the sample mean are:

$$\begin{aligned} \mu_{\overline{X}} &= \mu \\ \sigma_{\overline{X}}^2 &= \frac{\sigma^2}{n} \qquad \sigma_{\overline{X}} = \frac{\sigma}{\sqrt{n}} \\ \gamma_{1,\overline{X}} &= \frac{\gamma_1}{\sqrt{n}} \qquad \gamma_{2,\overline{X}} = \frac{\gamma_2}{n} \end{aligned} \qquad (4.26)$$

When the population is finite and of size M,

$$\begin{aligned} \mu_{\overline{x}}^{(M)} &= \mu \\ \sigma_{\overline{x}}^{2(M)} &= \frac{\sigma^2}{N} \frac{M-N}{M-1} \end{aligned} \qquad (4.27)$$

If the underlying population is *normal*, then the sample mean $\overline{X}$ is normally distributed.

4.3.3 Central limit theorem

Let $X_1, X_2, \ldots, X_n$ be a random sample from an infinite population with mean μ and variance σ^2. The limiting distribution of

$$Z = \frac{\overline{X} - \mu}{\sigma/\sqrt{n}} \qquad (4.28)$$

as $n \to \infty$ is the standard normal distribution. The limiting distribution of

$$T = \sum_{i=1}^{n} X_i \qquad (4.29)$$

as $n \to \infty$ is normal with mean $n\mu$ and variance $n\sigma^2$.

4.3.4 The law of large numbers

Let $X_1, X_2, \ldots, X_n$ be a random sample from an infinite population with mean μ and variance σ^2. For any positive constant c, the probability the sample mean is within c units of μ is at least $1 - \dfrac{\sigma^2}{nc^2}$:

$$\text{Prob}\left[\mu - c < \overline{X} < \mu + c\right] \geq 1 - \frac{\sigma^2}{nc^2}. \tag{4.30}$$

As $n \to \infty$ the probability approaches 1. (See Chebyshev inequality on page 38.)

4.4 FINITE POPULATION

Let $\{c_1, c_2, \ldots, c_N\}$ be a collection of numbers representing a finite population of size N and assume the sampling from this population is done without replacement. Let the random variable X_i be the ith observation selected from the population. The collection $X_1, X_2, \ldots, X_n$ is a random sample of size n from the finite population if the joint probability mass function for $X_1, X_2, \ldots, X_n$ is

$$p(x_1, x_2, \ldots, x_n) = \frac{1}{N(N-1)\cdots(N-n+1)}. \tag{4.31}$$

(1) The marginal probability distribution for the random variable X_i, $i = 1, 2, \ldots, n$ is

$$p_{X_i}(x_i) = \frac{1}{N} \quad \text{for} \quad x_i = c_1, c_2, \ldots, c_N. \tag{4.32}$$

(2) The mean and the variance of the finite population are

$$\mu = \sum_{i=1}^{N} c_i \frac{1}{N} \quad \text{and} \quad \sigma^2 = \sum_{i=1}^{N} (c_i - \mu)^2 \frac{1}{N}. \tag{4.33}$$

(3) The joint marginal probability mass function for any two random variables in the collection $X_1, X_2, \ldots, X_n$ is

$$p(x_i, x_j) = \frac{1}{N(N-1)}. \tag{4.34}$$

(4) The covariance between any two random variables in the collection $X_1, X_2, \ldots, X_n$ is

$$\text{Cov}\left[X_i, X_j\right] = -\frac{\sigma^2}{N-1}. \tag{4.35}$$

(5) Let $\overline{X}$ be the sample mean of the random sample of size n. The expected value and variance of $\overline{X}$ are

$$\text{E}\left[\overline{X}\right] = \mu \quad \text{and} \quad \text{Var}\left[\overline{X}\right] = \frac{\sigma^2}{n} \cdot \frac{N-n}{N-1}. \tag{4.36}$$

The quantity $(N-n)/(N-1)$ is the *finite population correction factor*.

4.5 THEOREMS

4.5.1 Theorems: the chi-square distribution

(1) Let Z be a standard normal random variable, then Z^2 has a chi-square distribution with 1 degree of freedom.

(2) Let $Z_1, Z_2, \ldots, Z_n$ be independent standard normal random variables. The random variable $Y = \sum_{i=1}^{n} Z_i^2$ has a chi-square distribution with n degrees of freedom.

(3) Let $X_1, X_2, \ldots, X_n$ be independent random variables such that X_i has a chi-square distribution with ν_i degrees of freedom. The random variable $Y = \sum_{i=1}^{n} X_i$ has a chi-square distribution with $\nu = \nu_1 + \nu_2 + \cdots + \nu_n$ degrees of freedom.

(4) Let U have a chi-square distribution with ν_1 degrees of freedom, U and V be independent, and $U + V$ have a chi-square distribution with $\nu > \nu_1$ degrees of freedom. The random variable V has a chi-sqaure distribution with $\nu - \nu_1$ degrees of freedom.

(5) Let $X_1, X_2, \ldots, X_n$ be a random sample from a normal population with mean μ and variance σ^2. Then

   (a) The sample mean, $\overline{X}$, and the sample variance, S^2, are independent, and

   (b) The random variable $\dfrac{(n-1)S^2}{\sigma^2}$ has a chi-square distribution with $n - 1$ degrees of freedom.

4.5.2 Theorems: the t distribution

(1) Let Z have a standard normal distribution, X have a chi-square distribution with ν degrees of freedom, and X and Z be independent. The random variable
$$T = \frac{Z}{\sqrt{X/\nu}} \qquad (4.37)$$
has a t distribution with ν degrees of freedom.

(2) Let $X_1, X_2, \ldots, X_n$ be a random sample from a normal population with mean μ and variance σ^2. The random variable
$$T = \frac{\overline{X} - \mu}{S/\sqrt{n}} \qquad (4.38)$$
has a t distribution with $n - 1$ degrees of freedom.

4.5.3 Theorems: the F distribution

(1) Let U have a chi-square distribution with ν_1 degrees of freedom, V have a chi-square distribution with ν_2 degrees of freedom, and U and V be

independent. The random variable

$$F = \frac{U/\nu_1}{V/\nu_2} \tag{4.39}$$

has an F distribution with ν_1 and ν_2 degrees of freedom.

(2) Let $X_1, X_2, \ldots, X_m$ and $Y_1, Y_2, \ldots, Y_n$ be random samples from normal populations with variances σ_X^2 and σ_Y^2, respectively. The random variable

$$F = \frac{S_X^2/\sigma_X^2}{S_y^2/\sigma_Y^2} \tag{4.40}$$

has an F distribution with $m-1$ and $n-1$ degrees of freedom.

(3) Let F_{α,ν_1,ν_2} be a critical value for the F distribution defined by $\text{Prob}\left[F \geq F_{\alpha,\nu_1,\nu_2}\right] = \alpha$. Then $F_{1-\alpha,\nu_1,\nu_2} = 1/F_{\alpha,\nu_2,\nu_1}$.

4.6 ORDER STATISTICS

4.6.1 Definition

Let $X_1, X_2, \ldots, X_n$ be independent continuous random variables with probability density function $f(x)$ and cumulative distribution function $F(x)$. The order statistic, $X_{(i)}$, $i = 1, 2, \ldots, n$, is a random variable defined to be the ith largest of the set $\{X_1, X_2, \ldots, X_n\}$. Therefore,

$$X_{(1)} \leq X_{(2)} \leq \cdots \leq X_{(n)} \tag{4.41}$$

and in particular

$$\begin{aligned} X_{(1)} &= \min\{X_1, X_2, \ldots, X_n\} \quad \text{and} \\ X_{(n)} &= \max\{X_1, X_2, \ldots, X_n\}. \end{aligned} \tag{4.42}$$

The cumulative distribution function for the ith order statistic is

$$F_{X_{(i)}}(x) = \text{Prob}\left[X_{(i)} \leq x\right] = \text{Prob}\left[i \text{ or more observations are} \leq x\right]$$

$$= \sum_{j=i}^{n} \binom{n}{j} [F(x)]^j [1 - F(x)]^{n-j} \tag{4.43}$$

and the probability density function is

$$\begin{aligned} f_{X_{(i)}}(x) &= n\binom{n-1}{i-1}[F(x)]^{i-1}[1-F(x)]^{n-i}f(x) \\ &= \frac{n!}{(i-1)!(n-i)!}[F(x)]^{i-1}f(x)[1-F(x)]^{n-i}. \end{aligned} \tag{4.44}$$

4.6. ORDER STATISTICS

4.6.2 The first order statistic

The probability density function, $f_{X_{(1)}}(x)$, and the cumulative distribution function, $F_{X_{(1)}}(x)$, for $X_{(1)}$ are

$$f_{X_{(1)}}(x) = n[1 - F(x)]^{n-1} f(x) \qquad F_{X_{(1)}}(x) = 1 - [1 - F(x)]^n. \qquad (4.45)$$

4.6.3 The nth order statistic

The probability density function, $f_{(n)}(x)$, and the cumulative distribution function, $F_{(n)}(x)$, for $X_{(n)}$ are

$$f_{X_{(n)}}(x) = n[F(x)]^{n-1} f(x) \qquad F_{X_{(n)}}(x) = [F(x)]^n. \qquad (4.46)$$

4.6.4 The median

If the number of observations is odd, the median is the middle observation when the observations are in numerical order. If the number of observations is even, the median is (arbitrarily) defined as the average of the middle two of the ordered observations.

$$\text{median} = \begin{cases} X_{(k)} & \text{if } n \text{ is odd and } n = 2k - 1 \\ \frac{1}{2}[X_{(k)} + X_{(k+1)}] & \text{if } n \text{ is even and } n = 2k \end{cases} \qquad (4.47)$$

4.6.5 Joint distributions

The joint density function for $X_{(1)}, X_{(2)}, \ldots, X_{(n)}$ is

$$g(x_1, x_2, \ldots, x_n) = n! f(x_1) f(x_2) \cdots f(x_n). \qquad (4.48)$$

The joint density function for the ith and j^{th} ($i < j$) order statistics is

$$\begin{aligned} f_{ij}(x, y) = &\frac{n!}{(i-1)!(j-i-1)!(n-j)!} f(x) f(y) \\ &\times [F(x)]^{i-1} [1 - F(y)]^{n-j} [F(y) - F(x)]^{j-i-1}. \end{aligned} \qquad (4.49)$$

The joint distribution function for $X_{(1)}$ and $X_{(n)}$ is

$$\begin{aligned} F_{1n}(x, y) &= \text{Prob}\left[X_{(1)} \leq x \text{ and } X_{(n)} \leq y\right] \\ &= \begin{cases} [F(y)]^n - [F(y) - F(x)]^n & \text{if } x \leq y \\ [F(y)]^n & \text{if } x > y \end{cases} \end{aligned} \qquad (4.50)$$

and the joint density function is

$$f_{1n}(x, y) = \begin{cases} n(n-1) f(x) f(y) [F(y) - F(x)]^{n-2} & \text{if } x \leq y \\ 0 & \text{if } x > y \end{cases} \qquad (4.51)$$

4.6.6 Midrange and range

The *midrange* is defined to be $A = \frac{1}{2}[X_{(1)} + X_{(n)}]$. Using $f_{1n}(x,y)$ for the joint density function of $X_{(1)}$ and $X_{(n)}$ results in

$$f_A(x) = 2\int_{-\infty}^{x} f_{1n}(t, 2x-t)\, dt$$
$$= 2n(n-1)\int_{-\infty}^{x} f(t)f(2x-t)\left[F(2x-t) - F(t)\right]^{n-2} dt \tag{4.52}$$

The *range* is the difference between the largest and smallest observations: $R = X_{(n)} - X_{(1)}$. The random variable R is used in the construction of *tolerance intervals*.

$$f_R(r) = \int_{-\infty}^{\infty} f_{1n}(t, t+r)\, dt$$
$$= \begin{cases} n(n-1)\int_{-\infty}^{\infty} f(t)f(t+r)[F(t+r) - F(t)]^{n-2}\, dt & \text{if } r > 0 \\ 0 & \text{if } r \leq 0 \end{cases} \tag{4.53}$$

4.6.7 Uniform distribution: order statistics

If X is uniformly distributed on the interval $[0,1]$ then the density function for $X_{(i)}$ is

$$f_i(x) = n\binom{n-1}{i-1} x^{i-1}(1-x)^{n-i}, \quad 0 \leq x \leq 1 \tag{4.54}$$

which is a beta distribution with parameters i and $n - i + 1$.

(1) $\mathrm{E}\left[X_{(i)}\right] = \int_0^1 f_k(t)\, dt = \dfrac{i}{n+1}$.

(2) The expected value of the largest of n observations is $\dfrac{n}{n+1}$.

(3) The expected value of the smallest of n observations is $\dfrac{1}{n+1}$.

(4) The density function of the midrange is

$$f_A(x) = \begin{cases} n2^{n-1} x^{n-1} & \text{if } 0 < x \leq \frac{1}{2} \\ n2^{n-1}(1-x)^{n-1} & \text{if } \frac{1}{2} \leq x < 1 \end{cases} \tag{4.55}$$

(5) The density function of the range is

$$f_R(r) = \begin{cases} n(n-1)(1-r)r^{n-2} & \text{if } 0 < r < 1 \\ 0 & \text{otherwise} \end{cases} \tag{4.56}$$

4.6. ORDER STATISTICS

4.6.7.1 Tolerance intervals

In many applications, we need to estimate an interval in which a certain proportion of the population lies, with given probability. A tolerance interval may be constructed using the results relating to order statistics and the range. A table of required sample sizes for varying ranges and probabilities is in the following table.

Example 4.27: Assume a sample is drawn from a uniform population. Find a sample size n such that at least 99% of the sample population, with probability .95, lies between the smallest and largest observations. This problem may be written as a probability statement:

$$\begin{aligned} 0.95 &= \text{Prob}\,[F(Z_n) - F(Z_1) > 0.99] \\ &= \text{Prob}\,[R > 0.99] \\ &= n(n-1) \int_{0.99}^{1} (1-r)r^{n-2}\, dr \\ &= 1 - (0.99)^{n-1}(0.01n + 0.99) \end{aligned} \qquad (4.57)$$

Solving this results in the value $n \approx 473$.

Tolerance intervals, uniform distribution

Probability	This fraction of the total population is within the range							
	0.500	0.750	0.900	0.950	0.975	0.990	0.995	0.999
0.500	3	7	17	34	67	168	336	1679
0.750	5	10	26	53	107	269	538	2692
0.900	6	14	38	77	154	388	777	3889
0.950	8	17	46	93	188	473	947	4742
0.975	9	20	54	109	221	555	1112	5570
0.990	10	24	64	130	263	661	1325	6636
0.995	11	26	71	145	294	740	1483	7427

For tolerance intervals for normal samples, see section 7.3.

4.6.8 Normal distribution: order statistics

When the $\{X_i\}$ come from a standard normal distribution, the $\{X_{(i)}\}$ are called *standard order statistics*.

4.6.8.1 Expected value of normal order statistics

The tables on pages 52–53 gives expected values of standard order statistics

$$\text{E}\,[X_{(i)}] = n \binom{n-1}{i-1} \int_{-\infty}^{\infty} t f(t) [F(t)]^{i-1} [1 - F(t)]^{n-i}\, dt \qquad (4.58)$$

when $f(x) = \dfrac{e^{-x^2/2}}{\sqrt{2\pi}}$ and $F(x) = \displaystyle\int_{-\infty}^{x} \dfrac{e^{-t^2/2}}{\sqrt{2\pi}}\, dt$. Missing values (indicated by a dash) may be obtained from $\text{E}\,[X_{(i)}] = -\text{E}\,[X_{(n-i+1)}]$.

Example 4.28: If an average person takes five intelligence tests (each test having a normal distribution with a mean of 100 and a standard deviation of 20), what is the expected value of the largest score?

Solution:

(S1) We need to obtain the expected value of the largest normal order statistic when $n = 5$.

(S2) Using $n = 5$ and $i = 5$ in the table below yields (use $j = 1$) $E\left[X_{(5)}\right]\Big|_{n=5} = 1.1630$.

(S3) The expected value of the largest score is $100 + (1.1630)(20) \approx 123$.

Expected value of the ith normal order statistic (use $j = n - i + 1$)

j		$n=2$	3	4	5	6	7	8	9
1		0.5642	0.8463	1.0294	1.1629	1.2672	1.3522	1.4236	1.4850
2		—	0.0000	0.2970	0.4950	0.6418	0.7574	0.8522	0.9323
3		—	—	—	0.0000	0.2015	0.3527	0.4728	0.5720
4		—	—	—	—	—	0.0000	0.1526	0.2745
5		—	—	—	—	—	—	—	0.0000

j	$n=10$	11	12	13	14	15	16	17	18	19
1	1.5388	1.5865	1.6292	1.6680	1.7034	1.7359	1.7660	1.7939	1.8200	1.8445
2	1.0014	1.0619	1.1157	1.1641	1.2079	1.2479	1.2848	1.3188	1.3504	1.3800
3	0.6561	0.7288	0.7929	0.8498	0.9011	0.9477	0.9903	1.0295	1.0657	1.0995
4	0.3757	0.4619	0.5368	0.6028	0.6618	0.7149	0.7632	0.8074	0.8481	0.8859
5	0.1227	0.2249	0.3122	0.3883	0.4556	0.5157	0.5700	0.6195	0.6648	0.7066
6	—	0.0000	0.1025	0.1905	0.2672	0.3353	0.3962	0.4513	0.5016	0.5477
7	—	—	—	0.0000	0.0882	0.1653	0.2337	0.2952	0.3508	0.4016
8	—	—	—	—	—	0.0000	0.0772	0.1459	0.2077	0.2637
9	—	—	—	—	—	—	—	0.0000	0.0688	0.1307
10	—	—	—	—	—	—	—	—	—	0.0000

j	$n=20$	21	22	23	24	25	26	27	28	29
1	1.8675	1.8892	1.9097	1.9292	1.9477	1.9653	1.9822	1.9983	2.0137	2.0285
2	1.4076	1.4336	1.4581	1.4813	1.5034	1.5243	1.5442	1.5632	1.5814	1.5988
3	1.1310	1.1605	1.1883	1.2145	1.2393	1.2628	1.2851	1.3064	1.3268	1.3462
4	0.9210	0.9538	0.9846	1.0136	1.0409	1.0668	1.0914	1.1147	1.1370	1.1582
5	0.7454	0.7816	0.8153	0.8470	0.8769	0.9051	0.9318	0.9571	0.9812	1.0042
6	0.5903	0.6298	0.6667	0.7012	0.7336	0.7641	0.7929	0.8202	0.8462	0.8709
7	0.4483	0.4915	0.5316	0.5690	0.6040	0.6369	0.6679	0.6973	0.7251	0.7515
8	0.3149	0.3620	0.4056	0.4461	0.4839	0.5193	0.5527	0.5841	0.6138	0.6420
9	0.1869	0.2384	0.2857	0.3296	0.3704	0.4086	0.4443	0.4780	0.5098	0.5398
10	0.0620	0.1183	0.1699	0.2175	0.2616	0.3026	0.3410	0.3770	0.4109	0.4430
11	—	0.0000	0.0564	0.1081	0.1558	0.2000	0.2413	0.2798	0.3160	0.3501
12	—	—	—	0.0000	0.0518	0.0995	0.1439	0.1852	0.2239	0.2602
13	—	—	—	—	—	0.0000	0.0478	0.0922	0.1336	0.1724
14	—	—	—	—	—	—	—	0.0000	0.0444	0.0859
15	—	—	—	—	—	—	—	—	—	0.0000

Expected value of the ith normal order statistic (use $j = n - i + 1$)

j	n = 30	31	32	33	34	35	36	37	38	39
1	2.0427	2.0564	2.0696	2.0824	2.0947	2.1066	2.1181	2.1292	2.1401	2.1505
2	1.6156	1.6316	1.6471	1.6620	1.6763	1.6902	1.7036	1.7165	1.7291	1.7413
3	1.3648	1.3827	1.3999	1.4164	1.4323	1.4476	1.4624	1.4768	1.4906	1.5040
4	1.1786	1.1980	1.2167	1.2347	1.2520	1.2686	1.2847	1.3002	1.3151	1.3296
5	1.0262	1.0472	1.0673	1.0866	1.1052	1.1230	1.1402	1.1568	1.1729	1.1884
6	0.8944	0.9169	0.9385	0.9591	0.9789	0.9979	1.0163	1.0339	1.0510	1.0674
7	0.7767	0.8007	0.8236	0.8456	0.8666	0.8868	0.9063	0.9250	0.9430	0.9604
8	0.6689	0.6944	0.7188	0.7420	0.7644	0.7857	0.8063	0.8261	0.8451	0.8634
9	0.5683	0.5954	0.6213	0.6460	0.6695	0.6921	0.7138	0.7346	0.7547	0.7740
10	0.4733	0.5020	0.5294	0.5555	0.5804	0.6043	0.6271	0.6490	0.6701	0.6904
11	0.3823	0.4129	0.4418	0.4694	0.4957	0.5208	0.5449	0.5679	0.5900	0.6113
12	0.2945	0.3268	0.3575	0.3867	0.4144	0.4409	0.4662	0.4904	0.5136	0.5359
13	0.2088	0.2432	0.2757	0.3065	0.3358	0.3637	0.3903	0.4157	0.4401	0.4635
14	0.1247	0.1613	0.1957	0.2283	0.2592	0.2886	0.3166	0.3433	0.3689	0.3934
15	0.0415	0.0804	0.1170	0.1515	0.1842	0.2151	0.2446	0.2727	0.2995	0.3252
16	—	0.0000	0.0389	0.0755	0.1101	0.1428	0.1739	0.2034	0.2316	0.2585
17	—	—	—	0.0000	0.0367	0.0713	0.1040	0.1351	0.1647	0.1929
18	—	—	—	—	—	0.0000	0.0346	0.0674	0.0986	0.1282
19	—	—	—	—	—	—	—	0.0000	0.0328	0.0640
20	—	—	—	—	—	—	—	—	—	0.0000

4.7 RANGE AND STUDENTIZED RANGE

4.7.1 Probability integral of the range

Let $\{X_1, X_2, \ldots, X_n\}$ denote a random sample of size n from a population with standard deviation σ, density function $f(x)$, and cumulative distribution function $F(x)$. Let $\{X_{(1)}, X_{(2)}, \ldots, X_{(n)}\}$ denote the same values in ascending order of magnitude. The sample range R is defined by

$$R = X_{(n)} - X_{(1)} \qquad (4.59)$$

In standardized form

$$W = \frac{R}{\sigma} = \frac{X_{(n)} - X_{(1)}}{\sigma} \qquad (4.60)$$

The probability that the range exceeds some value R, for a sample of size n, is (see equation (4.53))

$$\begin{aligned}
\text{Prob}\begin{bmatrix}\text{range exceeds } R \text{ for} \\ \text{a sample of size } n\end{bmatrix} &= \int_R^\infty f_R(r)\,dr \\
&= n\int_{-\infty}^\infty \Big[F(t+R) - F(t)\Big]^{n-1} f(t)\,dt
\end{aligned} \qquad (4.61)$$

4.7.2 Percentage points, studentized range

The standardized range is $W = R/\sigma$ as defined in the previous section. If the population standard deviation σ is replaced by the sample standard deviation

s (computed from another sample from the same population), then the studentized range Q is given by $Q = R/S$. Here, R is the range of the sample of size n and S is the independent of R and has ν degrees of freedom. The probability integral for the studentized range is given by

$$\text{Prob}\,[Q \leq q] = \text{Prob}\left[\frac{R}{S} \leq q\right] = \int_0^\infty \frac{2^{1-\nu/2}\nu^{\nu/2}s^{\nu-1}e^{-\nu s^2/2}f(qs)}{\Gamma(\nu/2)}\,ds \quad (4.62)$$

where f is the probability integral of the range for samples of size n.

The following tables provide values of the studentized range for the normal density function $f(x) = \dfrac{1}{\sqrt{2\pi}}e^{-x^2/2}$.

4.7. RANGE AND STUDENTIZED RANGE

Upper 1% points of the studentized range
The entries are $q_{.01}$ where $\text{Prob}\,[Q < q_{.01}] = .99$.

$\nu \backslash n$	2	3	4	5	6	7	8	9	10	11	12	13	14	15	16	17	18	19	20
1	77.75	129.44	147.54	170.27	188.43	202.60	215.08	225.53	234.69	242.85	250.15	255.43	261.60	238.57	243.92	248.69	253.42	257.88	262.10
2	13.58	18.33	22.09	24.66	26.66	28.28	29.68	30.87	30.63	31.76	32.78	33.71	34.57	35.36	36.10	36.79	37.44	38.05	38.63
3	8.10	10.54	12.18	13.35	13.96	14.86	15.63	16.29	16.88	17.40	17.87	18.31	18.70	19.07	19.41	19.73	20.04	20.32	20.71
4	6.33	8.11	9.20	9.75	10.46	11.03	11.52	11.94	12.31	12.64	12.94	13.22	13.47	13.70	13.92	14.13	14.32	14.50	14.67
5	5.64	6.98	7.82	8.28	8.81	9.23	9.59	9.91	10.18	10.43	10.65	10.86	11.05	11.22	11.38	11.53	11.67	11.81	11.94
6	5.23	6.34	7.05	7.44	7.87	8.22	8.51	8.77	8.99	9.19	9.37	9.54	9.69	9.83	9.96	10.09	10.20	10.31	10.42
7	4.94	5.93	6.55	6.91	7.28	7.57	7.83	8.05	8.38	8.56	8.72	8.87	9.01	9.13	9.25	9.36	9.47	9.57	9.66
8	4.75	5.65	6.12	6.54	6.87	7.13	7.48	7.69	7.87	8.04	8.18	8.32	8.44	8.56	8.66	8.77	8.86	8.95	9.03
9	4.60	5.44	5.89	6.27	6.57	6.92	7.14	7.33	7.50	7.65	7.79	7.91	8.03	8.13	8.23	8.33	8.42	8.50	8.58
10	4.49	5.20	5.71	6.07	6.35	6.68	6.88	7.06	7.22	7.36	7.49	7.60	7.71	7.81	7.91	7.99	8.08	8.15	8.23
11	4.40	5.09	5.57	5.91	6.17	6.48	6.68	6.85	7.00	7.13	7.25	7.36	7.46	7.56	7.65	7.73	7.81	7.88	7.95
12	4.33	5.00	5.46	5.78	6.03	6.33	6.51	6.67	6.82	6.95	7.06	7.17	7.26	7.35	7.44	7.52	7.59	7.66	7.73
13	4.27	4.93	5.37	5.68	5.92	6.20	6.38	6.53	6.67	6.80	6.90	7.01	7.10	7.19	7.27	7.34	7.42	7.48	7.55
14	4.22	4.86	5.29	5.59	5.82	6.09	6.26	6.41	6.55	6.67	6.77	6.87	6.96	7.05	7.12	7.20	7.27	7.33	7.39
15	4.18	4.81	5.22	5.51	5.74	6.00	6.17	6.31	6.44	6.56	6.66	6.76	6.84	6.93	7.00	7.07	7.14	7.20	7.26
16	4.08	4.76	5.16	5.45	5.67	5.92	6.08	6.23	6.35	6.47	6.57	6.66	6.74	6.82	6.90	6.97	7.03	7.09	7.15
17	4.05	4.72	5.12	5.39	5.61	5.79	6.01	6.15	6.27	6.38	6.48	6.57	6.66	6.73	6.81	6.87	6.94	6.99	7.05
18	4.03	4.69	5.07	5.35	5.56	5.73	5.95	6.08	6.20	6.31	6.41	6.50	6.58	6.65	6.72	6.79	6.85	6.91	6.97
19	4.01	4.66	5.03	5.30	5.51	5.68	5.89	6.03	6.14	6.25	6.34	6.43	6.51	6.58	6.65	6.72	6.78	6.84	6.89
20	3.99	4.63	5.00	5.27	5.47	5.64	5.84	5.97	6.09	6.19	6.29	6.37	6.45	6.52	6.59	6.66	6.71	6.77	6.82
25	3.92	4.52	4.87	5.12	5.32	5.48	5.61	5.73	5.89	5.99	6.07	6.15	6.23	6.29	6.36	6.42	6.47	6.52	6.57
30	3.87	4.45	4.79	5.03	5.22	5.37	5.50	5.61	5.71	5.80	5.94	6.01	6.08	6.14	6.20	6.26	6.31	6.36	6.41
40	3.82	4.37	4.70	4.93	5.10	5.25	5.37	5.48	5.57	5.66	5.73	5.80	5.87	5.93	5.98	6.03	6.08	6.13	6.17
60	3.76	4.29	4.60	4.82	4.99	5.13	5.25	5.35	5.44	5.52	5.60	5.66	5.72	5.78	5.83	5.88	5.93	5.97	6.01
120	3.71	4.20	4.50	4.71	4.87	5.00	5.12	5.21	5.30	5.37	5.44	5.50	5.56	5.61	5.66	5.70	5.75	5.79	5.82
1000	3.64	4.13	4.42	4.62	4.78	4.91	5.02	5.11	5.20	5.27	5.34	5.40	5.46	5.51	5.56	5.61	5.65	5.69	5.73

Upper 5% points of the studentized range

The entries are $q_{.05}$ where $\text{Prob}\,[Q < q_{.05}] = .95$.

ν	$n=2$	3	4	5	6	7	8	9	10	11	12	13	14	15	16	17	18	19	20
1	17.79	26.70	32.79	37.07	39.84	42.67	45.05	47.10	48.89	50.49	51.91	53.14	54.28	55.36	56.34	57.22	58.05	58.83	59.56
2	6.10	8.31	9.81	10.89	11.70	12.52	13.25	13.87	14.42	14.91	15.35	15.75	16.13	16.47	16.79	17.09	17.38	17.64	17.90
3	4.50	5.91	6.83	7.46	8.03	8.50	8.90	9.24	9.56	9.84	10.09	10.31	10.52	10.72	10.90	11.06	11.22	11.37	11.52
4	3.93	5.04	5.76	6.26	6.68	7.03	7.32	7.58	7.80	8.00	8.19	8.35	8.50	8.65	8.78	8.90	9.02	9.12	9.23
5	3.63	4.60	5.22	5.65	6.04	6.33	6.58	6.80	7.00	7.17	7.33	7.47	7.60	7.72	7.83	7.94	8.03	8.12	8.21
6	3.46	4.34	4.88	5.28	5.63	5.90	6.12	6.32	6.49	6.65	6.79	6.92	7.04	7.14	7.24	7.34	7.43	7.51	7.59
7	3.35	4.15	4.67	5.05	5.36	5.61	5.82	6.00	6.16	6.30	6.43	6.55	6.66	6.76	6.85	6.94	7.02	7.10	7.17
8	3.26	4.03	4.53	4.88	5.15	5.40	5.60	5.77	5.92	6.06	6.18	6.29	6.39	6.48	6.57	6.65	6.73	6.80	6.87
9	3.20	3.95	4.42	4.75	5.01	5.25	5.43	5.60	5.74	5.87	5.98	6.09	6.19	6.28	6.36	6.44	6.51	6.58	6.64
10	3.15	3.88	4.33	4.65	4.90	5.11	5.31	5.46	5.60	5.72	5.83	5.94	6.03	6.12	6.19	6.27	6.34	6.41	6.47
11	3.11	3.83	4.27	4.58	4.82	5.02	5.20	5.35	5.49	5.61	5.71	5.81	5.90	5.99	6.06	6.14	6.20	6.27	6.33
12	3.08	3.78	4.21	4.52	4.75	4.94	5.11	5.27	5.40	5.51	5.62	5.71	5.80	5.88	5.95	6.02	6.09	6.15	6.21
13	3.06	3.75	4.17	4.46	4.69	4.88	5.04	5.18	5.32	5.43	5.53	5.63	5.71	5.79	5.86	5.93	6.00	6.06	6.11
14	3.04	3.72	4.13	4.42	4.65	4.83	4.99	5.12	5.24	5.35	5.46	5.56	5.64	5.71	5.79	5.85	5.92	5.97	6.03
15	3.02	3.69	4.10	4.38	4.61	4.79	4.94	5.08	5.19	5.30	5.39	5.48	5.56	5.64	5.70	5.79	5.85	5.90	5.96
16	3.00	3.67	4.07	4.35	4.57	4.75	4.90	5.03	5.15	5.26	5.35	5.44	5.51	5.59	5.66	5.72	5.78	5.84	5.89
17	2.99	3.65	4.05	4.33	4.54	4.72	4.87	5.00	5.11	5.22	5.31	5.39	5.47	5.55	5.61	5.68	5.73	5.79	5.84
18	2.97	3.63	4.02	4.30	4.52	4.69	4.84	4.97	5.08	5.18	5.28	5.36	5.44	5.51	5.57	5.64	5.70	5.75	5.80
19	2.95	3.62	4.00	4.28	4.49	4.67	4.81	4.94	5.05	5.15	5.24	5.33	5.40	5.48	5.54	5.60	5.66	5.72	5.77
20	2.94	3.60	3.99	4.26	4.47	4.65	4.79	4.92	5.03	5.13	5.22	5.30	5.38	5.45	5.52	5.58	5.63	5.69	5.74
25	2.91	3.55	3.92	4.19	4.39	4.56	4.71	4.83	4.94	5.03	5.12	5.20	5.28	5.35	5.41	5.47	5.53	5.58	5.63
30	2.88	3.51	3.88	4.14	4.34	4.51	4.65	4.77	4.88	4.97	5.06	5.14	5.21	5.28	5.34	5.40	5.46	5.51	5.56
40	2.86	3.47	3.83	4.09	4.28	4.44	4.58	4.70	4.80	4.90	4.98	5.06	5.13	5.20	5.26	5.30	5.35	5.40	5.45
60	2.83	3.43	3.78	4.03	4.22	4.38	4.52	4.63	4.73	4.81	4.89	4.97	5.04	5.10	5.16	5.22	5.27	5.32	5.36
120	2.80	3.36	3.68	3.91	4.10	4.25	4.38	4.49	4.59	4.67	4.75	4.83	4.89	4.95	5.01	5.07	5.12	5.16	5.20
1000	2.77	3.35	3.68	3.92	4.11	4.25	4.37	4.48	4.58	4.67	4.75	4.82	4.88	4.95	5.00	5.06	5.11	5.15	5.20

Upper 10% points of the studentized range

The entries are $q_{.10}$ where $\text{Prob}[Q < q_{.10}] = .90$.

ν	$n=2$	3	4	5	6	7	8	9	10	11	12	13	14	15	16	17	18	19	20
1	8.94	13.43	16.37	18.43	20.12	21.49	22.63	23.61	24.47	25.23	25.91	26.52	26.81	27.46	28.06	28.62	29.15	29.65	30.18
2	4.13	5.73	6.77	7.54	8.14	8.63	9.05	9.41	9.73	10.01	10.26	10.49	10.70	10.89	11.07	11.24	11.40	11.54	11.63
3	3.33	4.47	5.20	5.74	6.16	6.51	6.81	7.06	7.29	7.49	7.67	7.83	7.98	8.12	8.25	8.37	8.48	8.58	8.68
4	3.01	3.98	4.59	5.03	5.39	5.68	5.92	6.14	6.33	6.49	6.64	6.78	6.91	7.02	7.13	7.23	7.33	7.41	7.50
5	2.85	3.72	4.26	4.66	4.98	5.24	5.46	5.65	5.81	5.96	6.10	6.22	6.33	6.44	6.53	6.62	6.71	6.79	6.86
6	2.75	3.56	4.07	4.44	4.73	4.97	5.17	5.34	5.50	5.64	5.76	5.87	5.98	6.07	6.16	6.25	6.32	6.40	6.46
7	2.68	3.45	3.93	4.28	4.55	4.78	4.97	5.14	5.28	5.41	5.53	5.64	5.73	5.82	5.91	5.99	6.06	6.13	6.19
8	2.63	3.37	3.83	4.17	4.43	4.65	4.83	4.99	5.12	5.25	5.36	5.46	5.56	5.64	5.72	5.80	5.87	5.93	6.00
9	2.59	3.32	3.76	4.08	4.34	4.54	4.72	4.87	5.01	5.13	5.23	5.33	5.42	5.50	5.58	5.65	5.72	5.79	5.84
10	2.56	3.27	3.70	4.02	4.26	4.47	4.64	4.78	4.91	5.03	5.13	5.23	5.32	5.40	5.47	5.54	5.61	5.67	5.73
11	2.54	3.23	3.66	3.97	4.20	4.40	4.57	4.71	4.84	4.95	5.05	5.15	5.23	5.31	5.38	5.45	5.51	5.57	5.63
12	2.52	3.20	3.62	3.92	4.16	4.35	4.51	4.65	4.78	4.89	4.99	5.08	5.16	5.24	5.31	5.37	5.44	5.49	5.55
13	2.51	3.18	3.59	3.89	4.12	4.30	4.46	4.60	4.72	4.83	4.93	5.02	5.10	5.18	5.25	5.31	5.37	5.43	5.48
14	2.49	3.16	3.56	3.85	4.08	4.27	4.42	4.56	4.68	4.79	4.88	4.97	5.05	5.12	5.19	5.26	5.32	5.37	5.43
15	2.48	3.14	3.54	3.83	4.05	4.24	4.39	4.52	4.64	4.75	4.84	4.93	5.01	5.08	5.15	5.21	5.27	5.32	5.38
16	2.47	3.12	3.52	3.81	4.03	4.21	4.36	4.49	4.61	4.71	4.81	4.89	4.97	5.04	5.11	5.17	5.23	5.28	5.33
17	2.46	3.11	3.50	3.78	4.00	4.18	4.33	4.46	4.58	4.68	4.77	4.86	4.93	5.01	5.07	5.13	5.19	5.24	5.30
18	2.45	3.10	3.49	3.77	3.98	4.16	4.31	4.44	4.55	4.66	4.75	4.83	4.91	4.98	5.04	5.10	5.16	5.21	5.26
19	2.44	3.09	3.47	3.75	3.97	4.14	4.29	4.42	4.53	4.63	4.72	4.80	4.88	4.95	5.01	5.07	5.13	5.18	5.23
20	2.43	3.08	3.46	3.74	3.95	4.12	4.27	4.40	4.51	4.61	4.70	4.78	4.86	4.92	4.99	5.05	5.10	5.16	5.21
25	2.41	3.04	3.42	3.68	3.89	4.06	4.20	4.32	4.43	4.53	4.62	4.69	4.77	4.83	4.89	4.95	5.00	5.06	5.10
30	2.40	3.02	3.39	3.65	3.85	4.02	4.16	4.28	4.38	4.47	4.56	4.64	4.71	4.77	4.83	4.89	4.94	4.99	5.04
40	2.38	2.99	3.35	3.61	3.80	3.96	4.10	4.22	4.32	4.41	4.49	4.56	4.63	4.69	4.75	4.81	4.86	4.91	4.95
60	2.36	2.96	3.31	3.56	3.76	3.91	4.04	4.16	4.25	4.34	4.42	4.49	4.56	4.62	4.68	4.73	4.78	4.82	4.87
120	2.34	2.93	3.28	3.52	3.71	3.86	3.99	4.10	4.19	4.28	4.35	4.42	4.49	4.54	4.60	4.65	4.69	4.74	4.78
1000	2.33	2.90	3.24	3.48	3.66	3.81	3.93	4.04	4.13	4.21	4.28	4.35	4.41	4.48	4.53	4.59	4.64	4.68	4.73

CHAPTER 5

Discrete Probability Distributions

This chapter presents some common discrete probability distributions along with their properties. Relevant numerical tables are also included.

Notation used throughout this chapter:

Probability mass function (pmf)	$p(x) = \text{Prob}[X = x]$
Mean	$\mu = \text{E}[X]$
Variance	$\sigma^2 = \text{E}\left[(X - \mu)^2\right]$
Coefficient of skewness	$\beta_1 = \text{E}\left[(X - \mu)^3\right]$
Coefficient of kurtosis	$\beta_2 = \text{E}\left[(X - \mu)^4\right]$
Moment generating function (mgf)	$m(t) = \text{E}\left[e^{tX}\right]$
Characteristic function (char function)	$\phi(t) = \text{E}\left[e^{itX}\right]$
Factorial moment generating function (fact mgf)	$P(t) = \text{E}\left[t^X\right]$

5.1 BERNOULLI DISTRIBUTION

A Bernoulli distribution is used to describe an experiment in which there are only two possible outcomes, typically a *success* or a *failure*. This type of experiment is called a *Bernoulli trial*, or simply a *trial*. The probability of a success is p and a sequence of Bernoulli trials is referred to as *repeated trials*.

5.1.1 Properties

pmf $\quad p(x) = \begin{cases} q & x = 0 \\ p & x = 1 \end{cases} \quad$ (or $p^x q^{1-x}$ for $x = 0, 1$)

$$0 \leq p \leq 1, \quad q = 1 - p$$

mean $\quad \mu = p$

variance	$\sigma^2 = pq$
skewness	$\beta_1 = \dfrac{1-2p}{\sqrt{pq}}$
kurtosis	$\beta_2 = 3 + \dfrac{1-6pq}{pq}$
mgf	$m(t) = q + pe^t$
char function	$\phi(t) = q + pe^{it}$
fact mgf	$P(t) = q + pt$

5.1.2 Variates

(1) Let $X_1, X_2, \ldots, X_n$ be independent, identically distributed (iid) Bernoulli random variables with probability of a success p. The random variable $Y = X_1 + X_2 + \cdots + X_n$ has a binomial distribution with parameters n and p.

5.2 BINOMIAL DISTRIBUTION

The binomial distribution is used to characterize the number of successes in n Bernoulli trials. It is used to model some very common experiments in which a sample of size n is taken from an infinite population such that each element is selected independently and has the same probability, p, of having a specified attribute.

5.2.1 Properties

pmf	$p(x) = \binom{n}{x} p^x q^{n-x} \quad x = 0, 1, 2, \ldots, n$
	$0 \leq p \leq 1, \quad q = 1 - p$
mean	$\mu = np$
variance	$\sigma^2 = npq$
skewness	$\beta_1 = \dfrac{1-2p}{\sqrt{npq}}$
kurtosis	$\beta_2 = 3 + \dfrac{1-6pq}{npq}$
mgf	$m(t) = (q + pe^t)^n$
char function	$\phi(t) = (q + pe^{it})^n$
fact mgf	$P(t) = (q + pt)^n$

5.2. BINOMIAL DISTRIBUTION

5.2.2 Variates

Let X be a binomial random variable with parameters n and p.

(1) If $n = 1$, then X is a Bernoulli random variable with probability of success p.

(2) As $n \to \infty$ if $np \geq 5$ and $n(1-p) \geq 5$, then X is approximately normal with parameters $\mu = np$ and $\sigma^2 = np(1-p)$.

(3) As $n \to \infty$ if $p < 0.1$ and $np < 10$, then X is approximately a Poisson random variable with parameter $\lambda = np$.

(4) Let $X_1, \ldots, X_k$ be independent, binomial random variables with parameters n_i and p, respectively. The random variable $Y = X_1+X_2+\cdots+X_k$ has a binomial distribution with parameters $n = n_1+n_2+\cdots+n_k$ and p.

5.2.3 Tables

The following tables only contain values of p up to $p = 1/2$. By symmetry (replacing p with $1-p$ and replacing x with $n-x$) values for $p > 1/2$ can be reduced to the present tables.

Example 5.29: A biased coin has a probability of heads of .75; what is the probability of obtaining 5 or more heads in 8 flips?

Solution:

(S1) The answer is given by looking in cumulative distribution tables with $n = 8$, $x = 5$, and $p = 0.75$.

(S2) Making the substitutions mentioned above this is the same as $n = 8$, $x = 3$, and $p = 0.25$.

(S3) This value is in the tables and is equal to 0.8862. Hence, 89% of the time 5 or more heads would be likely to occur.

Example 5.30: The probability a randomly selected home in Columbia County will lose power during a summer storm is .25. Suppose 14 homes in this county are selected at random. What is the probability exactly 4 homes will lose power, more than 6 will lose power, and between 2 and 7 (inclusive) will lose power?

Solution:

(S1) Let X be the number of homes (out of 14) that will lose power. The random variable X has a binomial distribution with parameters $n = 14$ and $p = 0.25$. Use the cumulative terms for the binomial distribution to answer each probability question.

(S2) $\text{Prob}[X = 4] = \text{Prob}[X \leq 4] - \text{Prob}[X \leq 3]$
$= 0.7415 - 0.5213 = 0.2202$

(S3) $\text{Prob}[X > 6] = 1 - \text{Prob}[X \leq 6] = 1 - 0.9617 = 0.0383$

(S4) $\text{Prob}[2 \leq X \leq 7] = \text{Prob}[X \leq 7] - \text{Prob}[X \leq 1]$
$= 0.9897 - 0.1010 = 0.8887$

Cumulative probability, Binomial distribution

n	x	p =0.05	0.10	0.15	0.20	0.25	0.30	0.40	0.50
2	0	0.9025	0.8100	0.7225	0.6400	0.5625	0.4900	0.3600	0.2500
	1	0.9975	0.9900	0.9775	0.9600	0.9375	0.9100	0.8400	0.7500
3	0	0.8574	0.7290	0.6141	0.5120	0.4219	0.3430	0.2160	0.1250
	1	0.9928	0.9720	0.9393	0.8960	0.8438	0.7840	0.6480	0.5000
	2	0.9999	0.9990	0.9966	0.9920	0.9844	0.9730	0.9360	0.8750
4	0	0.8145	0.6561	0.5220	0.4096	0.3164	0.2401	0.1296	0.0625
	1	0.9860	0.9477	0.8905	0.8192	0.7383	0.6517	0.4752	0.3125
	2	0.9995	0.9963	0.9880	0.9728	0.9492	0.9163	0.8208	0.6875
	3	1.0000	0.9999	0.9995	0.9984	0.9961	0.9919	0.9744	0.9375
5	0	0.7738	0.5905	0.4437	0.3277	0.2373	0.1681	0.0778	0.0313
	1	0.9774	0.9185	0.8352	0.7373	0.6328	0.5282	0.3370	0.1875
	2	0.9988	0.9914	0.9734	0.9421	0.8965	0.8369	0.6826	0.5000
	3	1.0000	0.9995	0.9978	0.9933	0.9844	0.9692	0.9130	0.8125
	4	1.0000	1.0000	0.9999	0.9997	0.9990	0.9976	0.9898	0.9688
6	0	0.7351	0.5314	0.3771	0.2621	0.1780	0.1177	0.0467	0.0156
	1	0.9672	0.8857	0.7765	0.6554	0.5339	0.4202	0.2333	0.1094
	2	0.9978	0.9841	0.9527	0.9011	0.8306	0.7443	0.5443	0.3438
	3	0.9999	0.9987	0.9941	0.9830	0.9624	0.9295	0.8208	0.6563
	4	1.0000	1.0000	0.9996	0.9984	0.9954	0.9891	0.9590	0.8906
	5	1.0000	1.0000	1.0000	0.9999	0.9998	0.9993	0.9959	0.9844
7	0	0.6983	0.4783	0.3206	0.2097	0.1335	0.0824	0.0280	0.0078
	1	0.9556	0.8503	0.7166	0.5767	0.4450	0.3294	0.1586	0.0625
	2	0.9962	0.9743	0.9262	0.8520	0.7564	0.6471	0.4199	0.2266
	3	0.9998	0.9973	0.9879	0.9667	0.9294	0.8740	0.7102	0.5000
	4	1.0000	0.9998	0.9988	0.9953	0.9871	0.9712	0.9037	0.7734
	5	1.0000	1.0000	0.9999	0.9996	0.9987	0.9962	0.9812	0.9375
	6	1.0000	1.0000	1.0000	1.0000	0.9999	0.9998	0.9984	0.9922
8	0	0.6634	0.4305	0.2725	0.1678	0.1001	0.0576	0.0168	0.0039
	1	0.9428	0.8131	0.6572	0.5033	0.3671	0.2553	0.1064	0.0352
	2	0.9942	0.9619	0.8948	0.7969	0.6785	0.5518	0.3154	0.1445
	3	0.9996	0.9950	0.9787	0.9437	0.8862	0.8059	0.5941	0.3633
	4	1.0000	0.9996	0.9971	0.9896	0.9727	0.9420	0.8263	0.6367
	5	1.0000	1.0000	0.9998	0.9988	0.9958	0.9887	0.9502	0.8555
	6	1.0000	1.0000	1.0000	0.9999	0.9996	0.9987	0.9915	0.9648
	7	1.0000	1.0000	1.0000	1.0000	1.0000	0.9999	0.9993	0.9961
9	0	0.6302	0.3874	0.2316	0.1342	0.0751	0.0403	0.0101	0.0019
	1	0.9288	0.7748	0.5995	0.4362	0.3003	0.1960	0.0705	0.0195
	2	0.9916	0.9470	0.8591	0.7382	0.6007	0.4628	0.2318	0.0898
	3	0.9994	0.9917	0.9661	0.9144	0.8343	0.7297	0.4826	0.2539
	4	1.0000	0.9991	0.9944	0.9804	0.9511	0.9012	0.7334	0.5000
	5	1.0000	0.9999	0.9994	0.9969	0.9900	0.9747	0.9006	0.7461
	6	1.0000	1.0000	1.0000	0.9997	0.9987	0.9957	0.9750	0.9102
	7	1.0000	1.0000	1.0000	1.0000	0.9999	0.9996	0.9962	0.9805
	8	1.0000	1.0000	1.0000	1.0000	1.0000	1.0000	0.9997	0.9980

5.2. BINOMIAL DISTRIBUTION

Cumulative probability, Binomial distribution

n	x	p =0.05	0.10	0.15	0.20	0.25	0.30	0.40	0.50
10	0	0.5987	0.3487	0.1969	0.1074	0.0563	0.0283	0.0060	0.0010
	1	0.9139	0.7361	0.5443	0.3758	0.2440	0.1493	0.0464	0.0107
	2	0.9885	0.9298	0.8202	0.6778	0.5256	0.3828	0.1673	0.0547
	3	0.9990	0.9872	0.9500	0.8791	0.7759	0.6496	0.3823	0.1719
	4	0.9999	0.9984	0.9901	0.9672	0.9219	0.8497	0.6331	0.3770
	5	1.0000	0.9999	0.9986	0.9936	0.9803	0.9526	0.8338	0.6230
	6	1.0000	1.0000	0.9999	0.9991	0.9965	0.9894	0.9452	0.8281
	7	1.0000	1.0000	1.0000	0.9999	0.9996	0.9984	0.9877	0.9453
	8	1.0000	1.0000	1.0000	1.0000	1.0000	0.9999	0.9983	0.9893
	9	1.0000	1.0000	1.0000	1.0000	1.0000	1.0000	0.9999	0.9990
11	0	0.5688	0.3138	0.1673	0.0859	0.0422	0.0198	0.0036	0.0005
	1	0.8981	0.6974	0.4922	0.3221	0.1971	0.1130	0.0302	0.0059
	2	0.9848	0.9104	0.7788	0.6174	0.4552	0.3127	0.1189	0.0327
	3	0.9984	0.9815	0.9306	0.8389	0.7133	0.5696	0.2963	0.1133
	4	0.9999	0.9972	0.9841	0.9496	0.8854	0.7897	0.5328	0.2744
	5	1.0000	0.9997	0.9973	0.9883	0.9657	0.9218	0.7535	0.5000
	6	1.0000	1.0000	0.9997	0.9980	0.9924	0.9784	0.9006	0.7256
	7	1.0000	1.0000	1.0000	0.9998	0.9988	0.9957	0.9707	0.8867
	8	1.0000	1.0000	1.0000	1.0000	0.9999	0.9994	0.9941	0.9673
	9	1.0000	1.0000	1.0000	1.0000	1.0000	1.0000	0.9993	0.9941
	10	1.0000	1.0000	1.0000	1.0000	1.0000	1.0000	1.0000	0.9995
12	0	0.5404	0.2824	0.1422	0.0687	0.0317	0.0138	0.0022	0.0002
	1	0.8816	0.6590	0.4435	0.2749	0.1584	0.0850	0.0196	0.0032
	2	0.9804	0.8891	0.7358	0.5584	0.3907	0.2528	0.0834	0.0193
	3	0.9978	0.9744	0.9078	0.7946	0.6488	0.4925	0.2253	0.0730
	4	0.9998	0.9957	0.9761	0.9274	0.8424	0.7237	0.4382	0.1938
	5	1.0000	0.9995	0.9954	0.9806	0.9456	0.8821	0.6652	0.3872
	6	1.0000	1.0000	0.9993	0.9961	0.9858	0.9614	0.8418	0.6128
	7	1.0000	1.0000	0.9999	0.9994	0.9972	0.9905	0.9427	0.8062
	8	1.0000	1.0000	1.0000	0.9999	0.9996	0.9983	0.9847	0.9270
	9	1.0000	1.0000	1.0000	1.0000	1.0000	0.9998	0.9972	0.9807
	10	1.0000	1.0000	1.0000	1.0000	1.0000	1.0000	0.9997	0.9968
	11	1.0000	1.0000	1.0000	1.0000	1.0000	1.0000	1.0000	0.9998
13	0	0.5133	0.2542	0.1209	0.0550	0.0238	0.0097	0.0013	0.0001
	1	0.8646	0.6213	0.3983	0.2336	0.1267	0.0637	0.0126	0.0017
	2	0.9755	0.8661	0.6920	0.5017	0.3326	0.2025	0.0579	0.0112
	3	0.9969	0.9658	0.8820	0.7473	0.5843	0.4206	0.1686	0.0461
	4	0.9997	0.9935	0.9658	0.9009	0.7940	0.6543	0.3530	0.1334
	5	1.0000	0.9991	0.9925	0.9700	0.9198	0.8346	0.5744	0.2905
	6	1.0000	0.9999	0.9987	0.9930	0.9757	0.9376	0.7712	0.5000
	7	1.0000	1.0000	0.9998	0.9988	0.9943	0.9818	0.9023	0.7095
	8	1.0000	1.0000	1.0000	0.9998	0.9990	0.9960	0.9679	0.8666
	9	1.0000	1.0000	1.0000	1.0000	0.9999	0.9993	0.9922	0.9539
	10	1.0000	1.0000	1.0000	1.0000	1.0000	0.9999	0.9987	0.9888
	11	1.0000	1.0000	1.0000	1.0000	1.0000	1.0000	0.9999	0.9983
	12	1.0000	1.0000	1.0000	1.0000	1.0000	1.0000	1.0000	0.9999

Cumulative probability, Binomial distribution

n	x	p =0.05	0.10	0.15	0.20	0.25	0.30	0.40	0.50
14	0	0.4877	0.2288	0.1028	0.0440	0.0178	0.0068	0.0008	0.0001
	1	0.8470	0.5846	0.3567	0.1979	0.1010	0.0475	0.0081	0.0009
	2	0.9699	0.8416	0.6479	0.4481	0.2811	0.1608	0.0398	0.0065
	3	0.9958	0.9559	0.8535	0.6982	0.5213	0.3552	0.1243	0.0287
	4	0.9996	0.9908	0.9533	0.8702	0.7415	0.5842	0.2793	0.0898
	5	1.0000	0.9985	0.9885	0.9562	0.8883	0.7805	0.4859	0.2120
	6	1.0000	0.9998	0.9978	0.9884	0.9617	0.9067	0.6925	0.3953
	7	1.0000	1.0000	0.9997	0.9976	0.9897	0.9685	0.8499	0.6047
	8	1.0000	1.0000	1.0000	0.9996	0.9979	0.9917	0.9417	0.7880
	9	1.0000	1.0000	1.0000	1.0000	0.9997	0.9983	0.9825	0.9102
	10	1.0000	1.0000	1.0000	1.0000	1.0000	0.9998	0.9961	0.9713
	11	1.0000	1.0000	1.0000	1.0000	1.0000	1.0000	0.9994	0.9935
	12	1.0000	1.0000	1.0000	1.0000	1.0000	1.0000	0.9999	0.9991
	13	1.0000	1.0000	1.0000	1.0000	1.0000	1.0000	1.0000	0.9999
15	0	0.4633	0.2059	0.0873	0.0352	0.0134	0.0047	0.0005	0.0000
	1	0.8290	0.5490	0.3186	0.1671	0.0802	0.0353	0.0052	0.0005
	2	0.9638	0.8159	0.6042	0.3980	0.2361	0.1268	0.0271	0.0037
	3	0.9945	0.9444	0.8227	0.6482	0.4613	0.2969	0.0905	0.0176
	4	0.9994	0.9873	0.9383	0.8358	0.6865	0.5155	0.2173	0.0592
	5	1.0000	0.9978	0.9832	0.9389	0.8516	0.7216	0.4032	0.1509
	6	1.0000	0.9997	0.9964	0.9819	0.9434	0.8689	0.6098	0.3036
	7	1.0000	1.0000	0.9994	0.9958	0.9827	0.9500	0.7869	0.5000
	8	1.0000	1.0000	0.9999	0.9992	0.9958	0.9848	0.9050	0.6964
	9	1.0000	1.0000	1.0000	0.9999	0.9992	0.9963	0.9662	0.8491
	10	1.0000	1.0000	1.0000	1.0000	0.9999	0.9993	0.9907	0.9408
	11	1.0000	1.0000	1.0000	1.0000	1.0000	0.9999	0.9981	0.9824
	12	1.0000	1.0000	1.0000	1.0000	1.0000	1.0000	0.9997	0.9963
	13	1.0000	1.0000	1.0000	1.0000	1.0000	1.0000	1.0000	0.9995
	14	1.0000	1.0000	1.0000	1.0000	1.0000	1.0000	1.0000	1.0000
16	0	0.4401	0.1853	0.0742	0.0282	0.0100	0.0033	0.0003	0.0000
	1	0.8108	0.5147	0.2839	0.1407	0.0635	0.0261	0.0033	0.0003
	2	0.9571	0.7893	0.5614	0.3518	0.1971	0.0994	0.0183	0.0021
	3	0.9930	0.9316	0.7899	0.5981	0.4050	0.2459	0.0651	0.0106
	4	0.9991	0.9830	0.9210	0.7983	0.6302	0.4499	0.1666	0.0384
	5	0.9999	0.9967	0.9765	0.9183	0.8104	0.6598	0.3288	0.1051
	6	1.0000	0.9995	0.9944	0.9733	0.9204	0.8247	0.5272	0.2273
	7	1.0000	0.9999	0.9989	0.9930	0.9729	0.9256	0.7161	0.4018
	8	1.0000	1.0000	0.9998	0.9985	0.9925	0.9743	0.8577	0.5982
	9	1.0000	1.0000	1.0000	0.9998	0.9984	0.9929	0.9417	0.7728
	10	1.0000	1.0000	1.0000	1.0000	0.9997	0.9984	0.9809	0.8949
	11	1.0000	1.0000	1.0000	1.0000	1.0000	0.9997	0.9951	0.9616
	12	1.0000	1.0000	1.0000	1.0000	1.0000	1.0000	0.9991	0.9894
	13	1.0000	1.0000	1.0000	1.0000	1.0000	1.0000	0.9999	0.9979
	14	1.0000	1.0000	1.0000	1.0000	1.0000	1.0000	1.0000	0.9997
	15	1.0000	1.0000	1.0000	1.0000	1.0000	1.0000	1.0000	1.0000

Cumulative probability, Binomial distribution

n	x	p=0.05	0.10	0.15	0.20	0.25	0.30	0.40	0.50
17	0	0.4181	0.1668	0.0631	0.0225	0.0075	0.0023	0.0002	0.0000
	1	0.7922	0.4818	0.2525	0.1182	0.0501	0.0193	0.0021	0.0001
	2	0.9497	0.7618	0.5198	0.3096	0.1637	0.0774	0.0123	0.0012
	3	0.9912	0.9174	0.7556	0.5489	0.3530	0.2019	0.0464	0.0064
	4	0.9988	0.9779	0.9013	0.7582	0.5739	0.3887	0.1260	0.0245
	5	0.9999	0.9953	0.9681	0.8943	0.7653	0.5968	0.2639	0.0717
	6	1.0000	0.9992	0.9917	0.9623	0.8929	0.7752	0.4478	0.1661
	7	1.0000	0.9999	0.9983	0.9891	0.9598	0.8954	0.6405	0.3145
	8	1.0000	1.0000	0.9997	0.9974	0.9876	0.9597	0.8011	0.5000
	9	1.0000	1.0000	1.0000	0.9995	0.9969	0.9873	0.9081	0.6855
	10	1.0000	1.0000	1.0000	0.9999	0.9994	0.9968	0.9652	0.8338
	11	1.0000	1.0000	1.0000	1.0000	0.9999	0.9993	0.9894	0.9283
	12	1.0000	1.0000	1.0000	1.0000	1.0000	0.9999	0.9975	0.9755
	13	1.0000	1.0000	1.0000	1.0000	1.0000	1.0000	0.9996	0.9936
	14	1.0000	1.0000	1.0000	1.0000	1.0000	1.0000	0.9999	0.9988
	15	1.0000	1.0000	1.0000	1.0000	1.0000	1.0000	1.0000	0.9999
	16	1.0000	1.0000	1.0000	1.0000	1.0000	1.0000	1.0000	1.0000
18	0	0.3972	0.1501	0.0537	0.0180	0.0056	0.0016	0.0001	0.0000
	1	0.7735	0.4503	0.2240	0.0991	0.0395	0.0142	0.0013	0.0001
	2	0.9419	0.7338	0.4797	0.2713	0.1353	0.0600	0.0082	0.0007
	3	0.9891	0.9018	0.7202	0.5010	0.3057	0.1646	0.0328	0.0038
	4	0.9984	0.9718	0.8794	0.7164	0.5187	0.3327	0.0942	0.0154
	5	0.9998	0.9936	0.9581	0.8671	0.7175	0.5344	0.2088	0.0481
	6	1.0000	0.9988	0.9882	0.9487	0.8610	0.7217	0.3743	0.1189
	7	1.0000	0.9998	0.9973	0.9837	0.9431	0.8593	0.5634	0.2403
	8	1.0000	1.0000	0.9995	0.9958	0.9807	0.9404	0.7368	0.4073
	9	1.0000	1.0000	0.9999	0.9991	0.9946	0.9790	0.8653	0.5927
	10	1.0000	1.0000	1.0000	0.9998	0.9988	0.9939	0.9424	0.7597
	11	1.0000	1.0000	1.0000	1.0000	0.9998	0.9986	0.9797	0.8811
	12	1.0000	1.0000	1.0000	1.0000	1.0000	0.9997	0.9942	0.9519
	13	1.0000	1.0000	1.0000	1.0000	1.0000	1.0000	0.9987	0.9846
	14	1.0000	1.0000	1.0000	1.0000	1.0000	1.0000	0.9998	0.9962
	15	1.0000	1.0000	1.0000	1.0000	1.0000	1.0000	1.0000	0.9993
	16	1.0000	1.0000	1.0000	1.0000	1.0000	1.0000	1.0000	0.9999
	17	1.0000	1.0000	1.0000	1.0000	1.0000	1.0000	1.0000	1.0000

Cumulative probability, Binomial distribution

n	x	p =0.05	0.10	0.15	0.20	0.25	0.30	0.40	0.50
19	0	0.3774	0.1351	0.0456	0.0144	0.0042	0.0011	0.0001	0.0000
	1	0.7547	0.4203	0.1985	0.0829	0.0310	0.0104	0.0008	0.0000
	2	0.9335	0.7054	0.4413	0.2369	0.1113	0.0462	0.0055	0.0004
	3	0.9868	0.8850	0.6842	0.4551	0.2631	0.1332	0.0230	0.0022
	4	0.9980	0.9648	0.8556	0.6733	0.4654	0.2822	0.0696	0.0096
	5	0.9998	0.9914	0.9463	0.8369	0.6678	0.4739	0.1629	0.0318
	6	1.0000	0.9983	0.9837	0.9324	0.8251	0.6655	0.3081	0.0835
	7	1.0000	0.9997	0.9959	0.9767	0.9225	0.8180	0.4878	0.1796
	8	1.0000	1.0000	0.9992	0.9933	0.9712	0.9161	0.6675	0.3238
	9	1.0000	1.0000	0.9999	0.9984	0.9911	0.9675	0.8139	0.5000
	10	1.0000	1.0000	1.0000	0.9997	0.9977	0.9895	0.9115	0.6762
	11	1.0000	1.0000	1.0000	1.0000	0.9995	0.9972	0.9648	0.8204
	12	1.0000	1.0000	1.0000	1.0000	0.9999	0.9994	0.9884	0.9165
	13	1.0000	1.0000	1.0000	1.0000	1.0000	0.9999	0.9969	0.9682
	14	1.0000	1.0000	1.0000	1.0000	1.0000	1.0000	0.9994	0.9904
	15	1.0000	1.0000	1.0000	1.0000	1.0000	1.0000	0.9999	0.9978
	16	1.0000	1.0000	1.0000	1.0000	1.0000	1.0000	1.0000	0.9996
	17	1.0000	1.0000	1.0000	1.0000	1.0000	1.0000	1.0000	1.0000
	18	1.0000	1.0000	1.0000	1.0000	1.0000	1.0000	1.0000	1.0000
20	0	0.3585	0.1216	0.0388	0.0115	0.0032	0.0008	0.0000	0.0000
	1	0.7358	0.3917	0.1756	0.0692	0.0243	0.0076	0.0005	0.0000
	2	0.9245	0.6769	0.4049	0.2061	0.0913	0.0355	0.0036	0.0002
	3	0.9841	0.8670	0.6477	0.4114	0.2252	0.1071	0.0160	0.0013
	4	0.9974	0.9568	0.8298	0.6297	0.4148	0.2375	0.0510	0.0059
	5	0.9997	0.9888	0.9327	0.8042	0.6172	0.4164	0.1256	0.0207
	6	1.0000	0.9976	0.9781	0.9133	0.7858	0.6080	0.2500	0.0577
	7	1.0000	0.9996	0.9941	0.9679	0.8982	0.7723	0.4159	0.1316
	8	1.0000	0.9999	0.9987	0.9900	0.9591	0.8867	0.5956	0.2517
	9	1.0000	1.0000	0.9998	0.9974	0.9861	0.9520	0.7553	0.4119
	10	1.0000	1.0000	1.0000	0.9994	0.9961	0.9829	0.8725	0.5881
	11	1.0000	1.0000	1.0000	0.9999	0.9991	0.9949	0.9435	0.7483
	12	1.0000	1.0000	1.0000	1.0000	0.9998	0.9987	0.9790	0.8684
	13	1.0000	1.0000	1.0000	1.0000	1.0000	0.9997	0.9935	0.9423
	14	1.0000	1.0000	1.0000	1.0000	1.0000	1.0000	0.9984	0.9793
	15	1.0000	1.0000	1.0000	1.0000	1.0000	1.0000	0.9997	0.9941
	16	1.0000	1.0000	1.0000	1.0000	1.0000	1.0000	1.0000	0.9987
	17	1.0000	1.0000	1.0000	1.0000	1.0000	1.0000	1.0000	0.9998
	18	1.0000	1.0000	1.0000	1.0000	1.0000	1.0000	1.0000	1.0000
	19	1.0000	1.0000	1.0000	1.0000	1.0000	1.0000	1.0000	1.0000

5.3 GEOMETRIC DISTRIBUTION

In a series of Bernoulli trials with probability of success p, a geometric random variable is the number of the trial on which the first success occurs. Hence, this is a waiting time distribution. The geometric distribution, sometimes

5.3. GEOMETRIC DISTRIBUTION

called the *Pascal distribution*, is often thought of as the discrete version of an exponential distribution.

5.3.1 Properties

pmf	$p(x) = pq^{x-1}$	$x = 1, 2, 3, \ldots,\quad 0 \le p \le 1,\quad q = 1-p$
mean	$\mu = 1/p$	
variance	$\sigma^2 = q/p^2$	
skewness	$\beta_1 = \dfrac{2-p}{\sqrt{q}}$	
kurtosis	$\beta_2 = \dfrac{p^2 + 6q}{q}$	
mgf	$m(t) = \dfrac{pe^t}{1 - qe^t}$	
char function	$\phi(t) = \dfrac{pe^{it}}{1 - qe^{it}}$	
fact mgf	$P(t) = \dfrac{p^t}{1 - qt}$	

5.3.2 Variates

Let $X_1, X_2, \ldots, X_n$ be independent, identically distributed geometric random variables with parameter p.

(1) The random variable $Y = X_1 + X_2 + \cdots + X_n$ has a negative binomial distribution with parameters n and p.

(2) The random variable $Y = \min(X_1, X_2, \ldots, X_n)$ has a geometric distribution with parameter p.

5.3.3 Tables

Example 5.31: When flipping a biased coin (so that heads occur only 30% of the time), what is the probability that the first head occurs on the 10th flip?

Solution:

(S1) Using the probability mass table below with $x = 10$ and $p = 0.3$ results in 0.0121.

(S2) Hence, this is likely to occur only about 1% of the time.

Example 5.32: The probability a randomly selected customer has the correct change when making a purchase at the local donut shop is 0.1. What is the probability the first person to have correct change will be the fifth customer? What is the probability the first person with correct change will be at least the sixth customer?

Solution:

(S1) Let X be the number of the first customer with correct change. The random variable X has a geometric distribution with parameter $p = 0.1$. Use the table for cumulative terms of the geometric probabilities to answer each question.

(S2) $\text{Prob}[X = 5] = 0.0656$

(S3) $\text{Prob}[X \geq 6] = 1 - \text{Prob}[X \leq 5]$
$= 1 - (0.1000 + 0.0900 + 0.0810 + 0.0729 + 0.656)$
$= 1 - 0.4095 = 0.5905$

Probability mass, Geometric distribution

x	$p = 0.1$	0.2	0.3	0.4	0.5	0.6	0.7	0.8	0.9
1	0.1000	0.2000	0.3000	0.4000	0.5000	0.6000	0.7000	0.8000	0.9000
2	0.0900	0.1600	0.2100	0.2400	0.2500	0.2400	0.2100	0.1600	0.0900
3	0.0810	0.1280	0.1470	0.1440	0.1250	0.0960	0.0630	0.0320	0.0090
4	0.0729	0.1024	0.1029	0.0864	0.0625	0.0384	0.0189	0.0064	0.0009
5	0.0656	0.0819	0.0720	0.0518	0.0313	0.0154	0.0057	0.0013	0.0001
6	0.0590	0.0655	0.0504	0.0311	0.0156	0.0061	0.0017	0.0003	
7	0.0531	0.0524	0.0353	0.0187	0.0078	0.0025	0.0005	0.0001	
8	0.0478	0.0419	0.0247	0.0112	0.0039	0.0010	0.0002		
9	0.0430	0.0336	0.0173	0.0067	0.0020	0.0004			
10	0.0387	0.0268	0.0121	0.0040	0.0010	0.0002			
11	0.0349	0.0215	0.0085	0.0024	0.0005	0.0001			
12	0.0314	0.0172	0.0059	0.0015	0.0002				
13	0.0282	0.0137	0.0042	0.0009	0.0001				
14	0.0254	0.0110	0.0029	0.0005	0.0001				
15	0.0229	0.0088	0.0020	0.0003					
20	0.0135	0.0029	0.0003						

Cumulative probability, Geometric distribution

x	$p = 0.1$	0.2	0.3	0.4	0.5	0.6	0.7	0.8	0.9
1	0.1000	0.2000	0.3000	0.4000	0.5000	0.6000	0.7000	0.8000	0.9000
2	0.1900	0.3600	0.5100	0.6400	0.7500	0.8400	0.9100	0.9600	0.9900
3	0.2710	0.4880	0.6570	0.7840	0.8750	0.9360	0.9730	0.9920	0.9990
4	0.3439	0.5904	0.7599	0.8704	0.9375	0.9744	0.9919	0.9984	0.9999
5	0.4095	0.6723	0.8319	0.9222	0.9688	0.9898	0.9976	0.9997	1
6	0.4686	0.7379	0.8824	0.9533	0.9844	0.9959	0.9993	0.9999	1
7	0.5217	0.7903	0.9176	0.9720	0.9922	0.9984	0.9998	1	
8	0.5695	0.8322	0.9424	0.9832	0.9961	0.9993	0.9999	1	
9	0.6126	0.8658	0.9596	0.9899	0.9980	0.9997	1		
10	0.6513	0.8926	0.9718	0.9940	0.9990	0.9999	1		
11	0.6862	0.9141	0.9802	0.9964	0.9995	1			
12	0.7176	0.9313	0.9862	0.9978	0.9998	1			
13	0.7458	0.9450	0.9903	0.9987	0.9999	1			
14	0.7712	0.9560	0.9932	0.9992	0.9999	1			
15	0.7941	0.9648	0.9953	0.9995	1				
20	0.8784	0.9885	0.9992	1					

5.4 HYPERGEOMETRIC DISTRIBUTION

In a finite population of size N suppose there are M successes (and $N - M$ failures). The hypergeometric distribution is used to describe the number of successes, X, in n trials (n observations drawn from the population). Unlike

5.4. HYPERGEOMETRIC DISTRIBUTION

a binomial distribution, the probability of a success does not remain constant from trial to trial.

5.4.1 Properties

$$\text{pmf} \quad p(x) = \frac{\binom{M}{x}\binom{N-M}{n-x}}{\binom{N}{n}} \quad x = 0, 1, \ldots, n \quad x \leq M$$

$$n - x \leq N - M, \quad n, M, N \in \mathcal{N}, \quad 1 \leq n \leq N$$

$$1 \leq M \leq N, \quad N = 1, 2, \ldots$$

mean $\quad \mu = n\dfrac{M}{N}$

variance $\quad \sigma^2 = \left(\dfrac{N-n}{N-1}\right) n\dfrac{M}{N}\left(1 - \dfrac{M}{N}\right)$

skewness $\quad \beta_1 = \dfrac{(N-2M)(N-2n)\sqrt{N-1}}{(N-2)\sqrt{nM(N-M)(N-n)}}$

kurtosis $\quad \beta_2 = \dfrac{N^2(N-1)}{(N-2)(N-3)nM(N-m)(N-n)} \times$

$$\left[N(N+1) - 6n(N-n) + 3\dfrac{M}{N^2}(N-M)\times\right.$$

$$\left.[N^2(n-2) - Nn^2 + 6n(N-n)]\right]$$

mgf $\quad m(t) = {}_2F_1(-n, -M; -N; 1 - e^t)$

char function $\quad \phi(t) = {}_2F_1(-n, -M; -N; 1 - e^{it})$

fact mgf $\quad P(t) = {}_2F_1(-n, -M; -N; 1 - t)$

where ${}_pF_q$ is the generalized hypergeometric function.

5.4.2 Variates

Let X be a hypergeometric random variable with parameters n, m, and N.

(1) As $N \to \infty$ if $n/N < 0.1$ then X is approximately a binomial random variable with parameters n and $p = M/N$.

(2) As n, M, and N all tend to infinity, if M/N is small then X has approximately a Poisson distribution with parameter $\lambda = nM/N$.

Example 5.33: A New York City transportation company has 10 taxis, 3 of which have broken meters. Suppose 4 taxis are selected at random. What is the probability exactly 1 will have a broken meter, fewer than 2 will have a broken meter, all will have working meters?

Solution:

(S1) Let X be the number of taxis selected with broken meters. The random variable X has a hypergeometric distribution with $N = 10$, $n = 4$, and $M = 3$.

(S2) $\text{Prob}\,[X = 1] = \dfrac{\binom{M}{1}\binom{N-M}{n-1}}{\binom{N}{n}} = \dfrac{\binom{3}{1}\binom{7}{3}}{\binom{10}{4}} = \dfrac{3 \cdot 35}{210} = 0.5$

(S3) $\text{Prob}\,[X = 0] = \dfrac{\binom{M}{0}\binom{N-M}{n}}{\binom{N}{n}} = \dfrac{\binom{3}{0}\binom{7}{4}}{\binom{10}{4}} = \dfrac{1 \cdot 35}{210} = 0.16667$

(S4) $\text{Prob}\,[X < 2] = \text{Prob}\,[X \leq 1] = \text{Prob}\,[X = 0] + \text{Prob}\,[X = 1] = 0.66667$

5.5 MULTINOMIAL DISTRIBUTION

The multinomial distribution is a generalization of the binomial distribution. Suppose there are n independent trials, and each trial results in exactly one of k possible distinct outcomes. For $i = 1, 2, \ldots, k$ let p_i be the probability that outcome i occurs on any given trial (with $\sum_{i=1}^{k} p_i = 1$). The multinomial random variable is the random vector $\mathbf{X} = [X_1, X_2, \ldots, X_k]^T$ where X_i is the number of times outcome i occurs.

5.5.1 Properties

$$\text{pmf} \quad p(x_1, x_2, \ldots, x_k) = n! \prod_{i=1}^{k} \frac{p_i^{x_i}}{x_i!}, \quad \sum_{i=1}^{k} x_i = n$$

$$\text{mean of } X_i \quad \mu_i = np_i$$

$$\text{variance of } X_i \quad \sigma_i^2 = np_i(1 - p_i)$$

$$\text{Cov}[X_i, X_j] \quad \sigma_{ij} = -np_ip_j, \quad i \neq j$$

$$\text{joint mgf} \quad m(t_1, t_2, \ldots, t_k) = (p_1 e^{t_1} + p_2 e^{t_2} + \cdots + p_k e^{t_k})^n$$

$$\text{joint char function} \quad \phi(t_1, t_2, \ldots, t_k) = (p_1 e^{it_1} + p_2 e^{it_2} + \cdots + p_k e^{it_k})^n$$

$$\text{joint fact mgf} \quad P(t_1, t_2, \ldots, t_k) = (p_1 t_1 + p_2 t_2 + \cdots + p_k t_k)^n$$

5.5.2 Variates

Let $\mathbf{X}$ be a multinomial random variable with parameters n (number of trials) and $p_1, p_2, \ldots, p_k$.

(1) The marginal distribution of X_i is binomial with parameters n and p_i.

(2) If $k = 2$ and $p_1 = p$, then the multinomial random variable corresponds to the binomial random variable with parameters n and p.

5.6 NEGATIVE BINOMIAL DISTRIBUTION

Consider a sequence of Bernoulli trials with probability of success p. The negative binomial distribution is used to describe the number of failures, X, before the nth success.

5.6. NEGATIVE BINOMIAL DISTRIBUTION

5.6.1 Properties

$$\text{pmf} \quad p(x) = \binom{x+n-1}{n-1} p^n q^x \quad x = 0, 1, 2, \ldots, \quad n = 1, 2, \ldots$$

$$0 \leq p \leq 1, \quad q = 1 - p$$

mean $\quad \mu = \dfrac{nq}{p}$

variance $\quad \sigma^2 = \dfrac{nq}{p^2}$

skewness $\quad \beta_1 = \dfrac{2-p}{\sqrt{nq}}$

kurtosis $\quad \beta_2 = 3 + \dfrac{p^2 + 6q}{nq}$

mgf $\quad m(t) = \left(\dfrac{p}{1 - qe^t}\right)^n$

char function $\quad \phi(t) = \left(\dfrac{p}{1 - qe^{it}}\right)^n$

fact mgf $\quad P(t) = \left(\dfrac{p}{1 - qt}\right)^n$

5.6.2 Variates

Let X be a negative binomial random variable with parameters n and p.

(1) If $n = 1$ then X is a geometric random variable with probability of success p.

(2) As $n \to \infty$ and $p \to 1$ with $n(1-p)$ held constant, X is approximately a Poisson random variable with $\lambda = n(1-p)$.

(3) Let $X_1, X_2, \ldots, X_k$ be independent negative binomial random variables with parameters n_i and p, respectively. The random variable $Y = X_1 + X_2 + \cdots + X_k$ has a negative binomial distribution with parameters $n = n_1 + n_2 + \cdots + n_k$ and p.

5.6.3 Tables

Example 5.34: Suppose a biased coin has probability of heads 0.3. What is the probability that the 5th head occurs after the 8th tail?

Solution:

(S1) Recognizing that $n = 5$ and $x = 8$ with $p = 0.3$ and $q = 1 - p = 0.7$, the probability is

$$\text{Prob}\,[X = 8] = \binom{5+8-1}{5-1}(0.3)^5(0.7)^8 = 495(0.3)^5(0.7)^8 = 0.0693$$

5.7 POISSON DISTRIBUTION

The Poisson, or *rare event*, distribution is completely described by a single parameter, λ. This distribution is used to model the number of successes, X, in a specified time interval or given region. It is assumed the numbers of successes occurring in different time intervals or regions are independent, the probability of a success in a time interval or region is very small and proportional to the length of the time interval or the size of the region, and the probability of more than one success during any one time interval or region is negligible.

5.7.1 Properties

$$\text{pmf} \quad p(x) = \frac{e^{-\lambda}\lambda^x}{x!} \quad x = 0, 1, 2, \ldots, \quad \lambda > 0$$

$$\text{mean} \quad \mu = \lambda$$

$$\text{variance} \quad \sigma^2 = \lambda$$

$$\text{skewness} \quad \beta_1 = 1/\sqrt{\lambda}$$

$$\text{kurtosis} \quad \beta_2 = 3 + (1/\lambda)$$

$$\text{mgf} \quad m(t) = \exp[\lambda(e^t - 1)]$$

$$\text{char function} \quad \phi(t) = \exp[\lambda(e^{it} - 1)]$$

$$\text{fact mgf} \quad P(t) = \exp[\lambda(t - 1)]$$

Note that the waiting time between Poisson arrivals is exponentially distributed.

5.7.2 Variates

Let X be a Poisson random variable with parameter λ.

(1) As $\lambda \to \infty$, X is approximately normal with parameters $\mu = \lambda$ and $\sigma^2 = \lambda$.

(2) Let $X_1, X_2, \ldots, X_n$ be independent Poisson random variables with parameters λ_i, respectively. The random variable $Y = X_1 + X_2 + \cdots + X_n$ has a Poisson distribution with parameter $\lambda = \lambda_1 + \lambda_2 + \cdots + \lambda_n$.

5.7.3 Tables

Example 5.35: The number of black bear sightings in Northeastern Pennsylvania during a given week has a Poisson distribution with $\lambda = 3$. For a randomly selected week, what is the probability of exactly 2 sightings, more than 5 sightings, between 4 and 7 sightings (inclusive)?

5.7. POISSON DISTRIBUTION

Solution:

(S1) Let X be the random variable representing the number of black bear sightings during any given week; X is Poisson with $\lambda = 3$. Use the table below to answer the probability questions.

(S2) $\text{Prob}\,[X = 2] = \text{Prob}\,[X \leq 2] - \text{Prob}\,[X \leq 1] = 0.423 - 0.199 = 0.224$

(S3) $\text{Prob}\,[X > 5] = 1 - \text{Prob}\,[X \leq 4] = 1 - 0.815 = 0.185$

(S4) $\text{Prob}\,[4 \leq X \leq 7] = \text{Prob}\,[X \leq 7] - \text{Prob}\,[X \leq 3] = .988 - .647 = .341$

Cumulative probability, Poisson distribution

λ	$x=0$	1	2	3	4	5	6	7	8	9
0.02	0.980	1.000								
0.04	0.961	0.999	1.000							
0.06	0.942	0.998	1.000							
0.08	0.923	0.997	1.000							
0.10	0.905	0.995	1.000							
0.15	0.861	0.990	1.000	1.000						
0.20	0.819	0.983	0.999	1.000						
0.25	0.779	0.974	0.998	1.000						
0.30	0.741	0.963	0.996	1.000						
0.35	0.705	0.951	0.995	1.000						
0.40	0.670	0.938	0.992	0.999	1.000					
0.45	0.638	0.925	0.989	0.999	1.000					
0.50	0.607	0.910	0.986	0.998	1.000					
0.55	0.577	0.894	0.982	0.998	1.000					
0.60	0.549	0.878	0.977	0.997	1.000					
0.65	0.522	0.861	0.972	0.996	0.999	1.000				
0.70	0.497	0.844	0.966	0.994	0.999	1.000				
0.75	0.472	0.827	0.960	0.993	0.999	1.000				
0.80	0.449	0.809	0.953	0.991	0.999	1.000				
0.85	0.427	0.791	0.945	0.989	0.998	1.000				
0.90	0.407	0.772	0.937	0.987	0.998	1.000				
0.95	0.387	0.754	0.929	0.984	0.997	1.000				
1.00	0.368	0.736	0.920	0.981	0.996	0.999	1.000			
1.1	0.333	0.699	0.900	0.974	0.995	0.999	1.000			
1.2	0.301	0.663	0.879	0.966	0.992	0.999	1.000			
1.3	0.273	0.627	0.857	0.957	0.989	0.998	1.000			
1.4	0.247	0.592	0.834	0.946	0.986	0.997	0.999	1.000		
1.5	0.223	0.558	0.809	0.934	0.981	0.996	0.999	1.000		
1.6	0.202	0.525	0.783	0.921	0.976	0.994	0.999	1.000		
1.7	0.183	0.493	0.757	0.907	0.970	0.992	0.998	1.000		
1.8	0.165	0.463	0.731	0.891	0.964	0.990	0.997	0.999	1.000	
1.9	0.150	0.434	0.704	0.875	0.956	0.987	0.997	0.999	1.000	
2.0	0.135	0.406	0.677	0.857	0.947	0.983	0.996	0.999	1.000	
2.2	0.111	0.355	0.623	0.819	0.927	0.975	0.993	0.998	1.000	
2.4	0.091	0.308	0.570	0.779	0.904	0.964	0.988	0.997	0.999	1.000
2.6	0.074	0.267	0.518	0.736	0.877	0.951	0.983	0.995	0.999	1.000

continued on next page

continued from previous page

λ	x =0	1	2	3	4	5	6	7	8	9
2.8	0.061	0.231	0.469	0.692	0.848	0.935	0.976	0.992	0.998	0.999
3.0	0.050	0.199	0.423	0.647	0.815	0.916	0.967	0.988	0.996	0.999
3.2	0.041	0.171	0.380	0.603	0.781	0.895	0.955	0.983	0.994	0.998
3.4	0.033	0.147	0.340	0.558	0.744	0.871	0.942	0.977	0.992	0.997
3.6	0.027	0.126	0.303	0.515	0.706	0.844	0.927	0.969	0.988	0.996
3.8	0.022	0.107	0.269	0.473	0.668	0.816	0.909	0.960	0.984	0.994
4.0	0.018	0.092	0.238	0.433	0.629	0.785	0.889	0.949	0.979	0.992
4.2	0.015	0.078	0.210	0.395	0.590	0.753	0.868	0.936	0.972	0.989
4.4	0.012	0.066	0.185	0.359	0.551	0.720	0.844	0.921	0.964	0.985
4.6	0.010	0.056	0.163	0.326	0.513	0.686	0.818	0.905	0.955	0.981
4.8	0.008	0.048	0.142	0.294	0.476	0.651	0.791	0.887	0.944	0.975
5.0	0.007	0.040	0.125	0.265	0.441	0.616	0.762	0.867	0.932	0.968
5.2	0.005	0.034	0.109	0.238	0.406	0.581	0.732	0.845	0.918	0.960
5.4	0.004	0.029	0.095	0.213	0.373	0.546	0.702	0.822	0.903	0.951
5.6	0.004	0.024	0.082	0.191	0.342	0.512	0.670	0.797	0.886	0.941
5.8	0.003	0.021	0.071	0.170	0.313	0.478	0.638	0.771	0.867	0.929
6.0	0.003	0.017	0.062	0.151	0.285	0.446	0.606	0.744	0.847	0.916
6.2	0.002	0.015	0.054	0.134	0.259	0.414	0.574	0.716	0.826	0.902
6.4	0.002	0.012	0.046	0.119	0.235	0.384	0.542	0.687	0.803	0.886
6.6	0.001	0.010	0.040	0.105	0.213	0.355	0.511	0.658	0.780	0.869
6.8	0.001	0.009	0.034	0.093	0.192	0.327	0.480	0.628	0.755	0.850
7.0	0.001	0.007	0.030	0.082	0.173	0.301	0.450	0.599	0.729	0.831
7.2	0.001	0.006	0.025	0.072	0.155	0.276	0.420	0.569	0.703	0.810
7.4	0.001	0.005	0.022	0.063	0.140	0.253	0.392	0.539	0.676	0.788
7.6	0.001	0.004	0.019	0.055	0.125	0.231	0.365	0.510	0.648	0.765
7.8	0.000	0.004	0.016	0.049	0.112	0.210	0.338	0.481	0.620	0.741
8.0	0.000	0.003	0.014	0.042	0.100	0.191	0.313	0.453	0.593	0.717
8.5	0.000	0.002	0.009	0.030	0.074	0.150	0.256	0.386	0.523	0.653
9.0	0.000	0.001	0.006	0.021	0.055	0.116	0.207	0.324	0.456	0.587
9.5	0.000	0.001	0.004	0.015	0.040	0.088	0.165	0.269	0.392	0.522
10.0	0.000	0.001	0.003	0.010	0.029	0.067	0.130	0.220	0.333	0.458
10.5	0.000	0.000	0.002	0.007	0.021	0.050	0.102	0.178	0.279	0.397
11.0	0.000	0.000	0.001	0.005	0.015	0.037	0.079	0.143	0.232	0.341
11.5	0.000	0.000	0.001	0.003	0.011	0.028	0.060	0.114	0.191	0.289
12.0	0.000	0.000	0.001	0.002	0.008	0.020	0.046	0.089	0.155	0.242
12.5	0.000	0.000	0.000	0.002	0.005	0.015	0.035	0.070	0.125	0.201
13.0	0.000	0.000	0.000	0.001	0.004	0.011	0.026	0.054	0.100	0.166
13.5	0.000	0.000	0.000	0.001	0.003	0.008	0.019	0.042	0.079	0.135
14.0	0.000	0.000	0.000	0.001	0.002	0.005	0.014	0.032	0.062	0.109
14.5	0.000	0.000	0.000	0.000	0.001	0.004	0.011	0.024	0.048	0.088
15.0	0.000	0.000	0.000	0.000	0.001	0.003	0.008	0.018	0.037	0.070

5.8 RECTANGULAR (DISCRETE UNIFORM) DISTRIBUTION

A general rectangular distribution is used to describe a random variable, X, that can assume n different values with equal probabilities. In the special case presented here, we assume the random variable can assume the first n positive integers.

5.8.1 Properties

pmf	$p(x) = 1/n, \quad x = 1, 2, \ldots, n, \quad n \in \mathcal{N}$
mean	$\mu = (n+1)/2$
variance	$\sigma^2 = (n^2 - 1)/12$
skewness	$\beta_1 = 0$
kurtosis	$\beta_2 = \dfrac{3}{5}\left(3 - \dfrac{4}{n^2 - 1}\right)$
mgf	$m(t) = \dfrac{e^t(1 - e^{nt})}{n(1 - e^t)}$
char function	$\phi(t) = \dfrac{e^{it}(1 - e^{nit})}{n(1 - e^{it})}$
fact mgf	$P(t) = \dfrac{t(1 - t^n)}{n(1 - t)}$

Example 5.36: A new family game has a special 12-sided numbered die, manufactured so that each side is equally likely to occur. Find the mean and variance of the number rolled, and the probability of rolling a 2, 3, or 12.

Solution:

(S1) Let X be the number on the side facing up; X has a discrete uniform distribution with $n = 12$.

(S2) Using the properties given above:
$$\mu = (n+1)/2 = (12+1)/2 = {}^{13}\!/_2 = 6.5$$
$$\sigma^2 = (n^2 - 1)/12 = (12^2 - 1)/12 = {}^{143}\!/_{12} = 11.9167$$

(S3) $\text{Prob}\,[X = 2, 3, 12] = \dfrac{1}{12} + \dfrac{1}{12} + \dfrac{1}{12} = \dfrac{3}{12} = 0.25$

CHAPTER 6

Continuous Probability Distributions

This chapter presents some common continuous probability distributions along with their properties. Relevant numerical tables are also included.

Notation used throughout this chapter:

Probability density function (pdf)	$f(x)$	$\text{Prob}[a \leq X \leq b] = \int_a^b f(x)\,dx$
Cumulative distrib function (cdf)		$F(x) = \text{Prob}[X \leq x] = \int_{-\infty}^x f(x)\,dx$
Mean		$\mu = \text{E}[X]$
Variance		$\sigma^2 = \text{E}\left[(X-\mu)^2\right]$
Coefficient of skewness		$\beta_1 = \text{E}\left[(X-\mu)^3\right]/\sigma^3$
Coefficient of kurtosis		$\beta_2 = \text{E}\left[(X-\mu)^4\right]/\sigma^4$
Moment generating function (mgf)		$m(t) = \text{E}\left[e^{tX}\right]$
Characteristic function (char function)		$\phi(t) = \text{E}\left[e^{itX}\right]$

where $\Gamma(x)$ is the gamma function, $B(a,b)$ is the beta function, and ${}_pF_q$ is the generalized hypergeometric function.

The rth moment about the origin is

$$\mu'_r = \frac{\Gamma(\alpha+\beta)\Gamma(\alpha+r)}{\Gamma(\alpha)\Gamma(\alpha+\beta+r)} \tag{6.1}$$

6.0.2 Probability density function

If $\alpha < 1$ and $\beta < 1$ the probability density function is "U" shaped. If the product $(\alpha-1)(\beta-1) < 0$ the probability density function is "J" shaped. Let $f(x;\alpha,\beta)$ denote the probability density function for a beta random variable with parameters α and β. If both $\alpha > 1$ and $\beta > 1$ then $f(x;\alpha,\beta)$ and $f(x;\beta,\alpha)$ are symmetric with respect to the line $x = .5$.

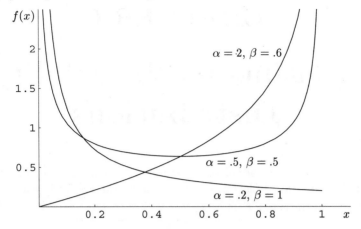

Figure 6.1: Probability density functions for a beta random variable, various shape parameters.

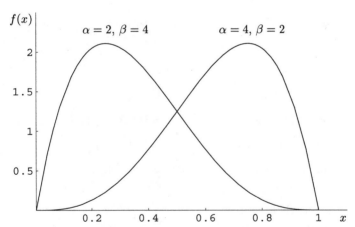

Figure 6.2: Probability density functions for a beta random variable, example of symmetry.

6.0.3 Related distributions

Let X be a beta random variable with parameters α and β.

(1) If $\alpha = \beta = 1/2$, then X is an arcsin random variable.

(2) If $\alpha = \beta = 1$, then X is a uniform random variable with parameters $a = 0$ and $b = 1$.

(3) If $\beta = 1$, then X is a power function random variable with parameters $b = 1$ and $c = \alpha$.

(4) As α and β tend to infinity such that α/β is constant, X tends to a standard normal random variable.

6.1 CAUCHY DISTRIBUTION

6.1.1 Properties

pdf	$f(x) = \dfrac{1}{b\pi\left(1+\left(\frac{x-a}{b}\right)^2\right)}$, $\quad x \in \mathcal{R},\ a \in \mathcal{R},\ b > 0$
mean	$\mu = $ does not exist
variance	$\sigma^2 = $ does not exist
skewness	$\beta_1 = $ does not exist
kurtosis	$\beta_2 = $ does not exist
mgf	$m(t) = $ does not exist
char function	$\phi(t) = e^{ait-b|t|}$

6.1.2 Probability density function

The probability density function for a Cauchy random variable is unimodal and symmetric about the parameter a. The tails are heavier than those of a normal random variable.

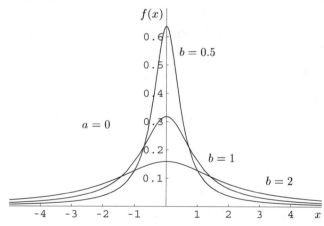

Figure 6.3: Probability density functions for a Cauchy random variable.

6.1.3 Related distributions

Let X be a Cauchy random variable with parameters a and b.

(1) If $a = 0$ and $b = 1$ then X is a standard Cauchy random variable.

(2) The random variable $1/X$ is also a Cauchy random variable with parameters $a/(a^2+b^2)$ and $b/(a^2+b^2)$.

(3) Let X_i (for $i = 1, 2, \ldots, n$) be independent, Cauchy random variables with parameters a_i and b_i, respectively. The random variable $Y = X_1 + X_2 + \cdots + X_n$ has a Cauchy distribution with parameters $a = a_1 + a_2 + \cdots + a_n$ and $b = b_1 + b_2 + \cdots + b_n$.

6.2 CHI-SQUARE DISTRIBUTION

6.2.1 Properties

$$\text{pdf} \quad f(x) = \frac{e^{-x/2} x^{(\nu/2)-1}}{2^{\nu/2} \Gamma(\nu/2)}, \quad x \geq 0, \; \nu \in \mathcal{N}$$

$$\text{mean} \quad \mu = \nu$$

$$\text{variance} \quad \sigma^2 = 2\nu$$

$$\text{skewness} \quad \beta_1 = 2\sqrt{2/\nu}$$

$$\text{kurtosis} \quad \beta_2 = 3 + \frac{12}{\nu}$$

$$\text{mgf} \quad m(t) = (1 - 2t)^{-\nu/2}, \quad t < 1/2$$

$$\text{char function} \quad \phi(t) = (1 - 2it)^{-\nu/2}$$

where $\Gamma(x)$ is the gamma function (see page 204).

A chi-square(χ^2) distribution is completely characterized by the parameter ν, the *degrees of freedom*.

6.2.2 Probability density function

The probability density function for a chi-square random variable is positively skewed. As ν tends to infinity, the density function becomes more bell-shaped and symmetric.

6.2.3 Related distributions

(1) If X is a chi-square random variable with $\nu = 2$, then X is an exponential random variable with $\lambda = 1/2$.

(2) If X_1 and X_2 are independent chi-square random variables with parameters ν_1 and ν_2, then the random variable $(X_1/\nu_1)/(X_2/\nu_2)$ has an F distribution with ν_1 and ν_2 degrees of freedom.

(3) If X_1 and X_2 are independent chi-square random variables with parameters $\nu_1 = \nu_2 = \nu$, the random variable

$$Y = \frac{\sqrt{\nu}}{2} \frac{X_1 - X_2}{\sqrt{X_1 X_2}} \tag{6.2}$$

6.2. CHI-SQUARE DISTRIBUTION

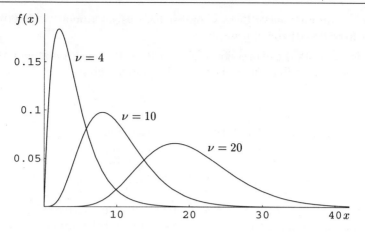

Figure 6.4: Probability density functions for a chi-square random variable.

has a t distribution with ν degrees of freedom.

(4) Let X_i (for $i = 1, 2, \ldots, n$) be independent chi-square random variables with parameters ν_i. The random variable $Y = X_1 + X_2 + \cdots + X_n$ has a chi-square distribution with $\nu = \nu_1 + \nu_2 + \cdots + \nu_n$ degrees of freedom.

(5) If X is a chi-square random variable with ν degrees of freedom, the random variable $\sqrt{X}$ has a **chi distribution** with parameter ν.
Properties of a chi random variable:

$$\text{pdf} \quad f(x) = \frac{x^{n-1} e^{-x^2/2}}{2^{(n/2)-1} \Gamma(n/2)}, \quad x \geq 0, \; n \in \mathcal{N}$$

$$\text{mean} \quad \mu = \frac{\Gamma\left(\frac{n+1}{2}\right)}{\Gamma\left(\frac{n}{2}\right)}$$

$$\text{variance} \quad \sigma^2 = \frac{\Gamma\left(\frac{n+2}{2}\right)}{\Gamma\left(\frac{n}{2}\right)} - \left[\frac{\Gamma\left(\frac{n+1}{2}\right)}{\Gamma\left(\frac{n}{2}\right)}\right]^2$$

where $\Gamma(x)$ is the gamma function (see page 204).
If X is a chi random variable with parameter $n = 2$, then X is a Rayleigh random variable with $\sigma = 1$.

6.2.4 Critical values for chi-square distribution

The following tables give values of $\chi^2_{\alpha,\nu}$ such that

$$1 - \alpha = F\left(\chi^2_{\alpha,\nu}\right) = \int_0^{\chi^2_{\alpha,\nu}} \frac{1}{2^{\nu/2} \Gamma(\nu/2)} x^{(\nu-2)/2} e^{-x/2} \, dx \quad (6.3)$$

where ν, the number of degrees of freedom, varies from 1 to 10,000 and α varies from 0.0001 to 0.9999.

(a) For $\nu > 30$, the expression $\sqrt{2\chi^2} - \sqrt{2\nu - 1}$ is approximately a standard normal distribution. Hence, $\chi^2_{\alpha,\nu}$ is approximately

$$\chi^2_{\alpha,\nu} \approx \frac{1}{2}\left[z_\alpha + \sqrt{2\nu - 1}\right]^2 \qquad \text{for } \nu \gg 1 \qquad (6.4)$$

(b) For even values of ν, $F\left(\chi^2_{\alpha,\nu}\right)$ can be written as

$$1 - F\left(\chi^2_{\alpha,\nu}\right) = \sum_{x=0}^{x'-1} \frac{e^{-\lambda}\lambda^x}{x!} \qquad (6.5)$$

with $\lambda = \chi^2_{\alpha,\nu}/2$ and $x' = \nu/2$. Hence, the cumulative chi-square distribution is related to the cumulative Poisson distribution.

Example 6.37: Use the table on page 83 to find the values $\chi^2_{.99,36}$ and $\chi^2_{.05,20}$.

Solution:

(S1) The left-hand column of the table on page 83 contains entries for the number of degrees of freedom and the top row lists values for α. The intersection of the ν degrees of freedom row and the α column contains $\chi^2_{\alpha,\nu}$ such that Prob $\left[\chi^2 \geq \chi^2_{\alpha,\nu}\right] = \alpha$.

(S2) $\chi^2_{.99,36} = 19.2327 \implies$ Prob $\left[\chi^2 \geq 19.2327\right] = .99$
$\chi^2_{.05,20} = 31.4104 \implies$ Prob $\left[\chi^2 \geq 31.4104\right] = .05$

6.2. CHI-SQUARE DISTRIBUTION

Critical values for the chi–square distribution $\chi^2_{\alpha,\nu}$.

ν	.9999	.9995	.999	α .995	.99	.975	.95	.90
1	$.0^7157$	$.0^6393$	$.0^5157$	$.0^4393$	.0002	.0010	.0039	.0158
2	.0002	.0010	.0020	.0100	.0201	.0506	.1026	.2107
3	.0052	.0153	.0243	.0717	.1148	.2158	.3518	.5844
4	.0284	.0639	.0908	.2070	.2971	.4844	.7107	1.0636
5	.0822	.1581	.2102	.4117	.5543	.8312	1.1455	1.6103
6	.1724	.2994	.3811	.6757	.8721	1.2373	1.6354	2.2041
7	.3000	.4849	.5985	.9893	1.2390	1.6899	2.1673	2.8331
8	.4636	.7104	.8571	1.3444	1.6465	2.1797	2.7326	3.4895
9	.6608	.9717	1.1519	1.7349	2.0879	2.7004	3.3251	4.1682
10	.8889	1.2650	1.4787	2.1559	2.5582	3.2470	3.9403	4.8652
11	1.1453	1.5868	1.8339	2.6032	3.0535	3.8157	4.5748	5.5778
12	1.4275	1.9344	2.2142	3.0738	3.5706	4.4038	5.2260	6.3038
13	1.7333	2.3051	2.6172	3.5650	4.1069	5.0088	5.8919	7.0415
14	2.0608	2.6967	3.0407	4.0747	4.6604	5.6287	6.5706	7.7895
15	2.4082	3.1075	3.4827	4.6009	5.2293	6.2621	7.2609	8.5468
16	2.7739	3.5358	3.9416	5.1422	5.8122	6.9077	7.9616	9.3122
17	3.1567	3.9802	4.4161	5.6972	6.4078	7.5642	8.6718	10.0852
18	3.5552	4.4394	4.9048	6.2648	7.0149	8.2307	9.3905	10.8649
19	3.9683	4.9123	5.4068	6.8440	7.6327	8.9065	10.1170	11.6509
20	4.3952	5.3981	5.9210	7.4338	8.2604	9.5908	10.8508	12.4426
21	4.8348	5.8957	6.4467	8.0337	8.8972	10.2829	11.5913	13.2396
22	5.2865	6.4045	6.9830	8.6427	9.5425	10.9823	12.3380	14.0415
23	5.7494	6.9237	7.5292	9.2604	10.1957	11.6886	13.0905	14.8480
24	6.2230	7.4527	8.0849	9.8862	10.8564	12.4012	13.8484	15.6587
25	6.7066	7.9910	8.6493	10.5197	11.5240	13.1197	14.6114	16.4734
26	7.1998	8.5379	9.2221	11.1602	12.1981	13.8439	15.3792	17.2919
27	7.7019	9.0932	9.8028	11.8076	12.8785	14.5734	16.1514	18.1139
28	8.2126	9.6563	10.3909	12.4613	13.5647	15.3079	16.9279	18.9392
29	8.7315	10.2268	10.9861	13.1211	14.2565	16.0471	17.7084	19.7677
30	9.2581	10.8044	11.5880	13.7867	14.9535	16.7908	18.4927	20.5992
31	9.7921	11.3887	12.1963	14.4578	15.6555	17.5387	19.2806	21.4336
32	10.3331	11.9794	12.8107	15.1340	16.3622	18.2908	20.0719	22.2706
33	10.8810	12.5763	13.4309	15.8153	17.0735	19.0467	20.8665	23.1102
34	11.4352	13.1791	14.0567	16.5013	17.7891	19.8063	21.6643	23.9523
35	11.9957	13.7875	14.6878	17.1918	18.5089	20.5694	22.4650	24.7967
36	12.5622	14.4012	15.3241	17.8867	19.2327	21.3359	23.2686	25.6433
37	13.1343	15.0202	15.9653	18.5858	19.9602	22.1056	24.0749	26.4921
38	13.7120	15.6441	16.6112	19.2889	20.6914	22.8785	24.8839	27.3430
39	14.2950	16.2729	17.2616	19.9959	21.4262	23.6543	25.6954	28.1958
40	14.8831	16.9062	17.9164	20.7065	22.1643	24.4330	26.5093	29.0505

Critical values for the chi–square distribution $\chi^2_{\alpha,\nu}$.

ν	.9999	.9995	.999	α .995	.99	.975	.95	.90
41	15.48	17.54	18.58	21.42	22.91	25.21	27.33	29.91
42	16.07	18.19	19.24	22.14	23.65	26.00	28.14	30.77
43	16.68	18.83	19.91	22.86	24.40	26.79	28.96	31.63
44	17.28	19.48	20.58	23.58	25.15	27.57	29.79	32.49
45	17.89	20.14	21.25	24.31	25.90	28.37	30.61	33.35
46	18.51	20.79	21.93	25.04	26.66	29.16	31.44	34.22
47	19.13	21.46	22.61	25.77	27.42	29.96	32.27	35.08
48	19.75	22.12	23.29	26.51	28.18	30.75	33.10	35.95
49	20.38	22.79	23.98	27.25	28.94	31.55	33.93	36.82
50	21.01	23.46	24.67	27.99	29.71	32.36	34.76	37.69
60	27.50	30.34	31.74	35.53	37.48	40.48	43.19	46.46
70	34.26	37.47	39.04	43.28	45.44	48.76	51.74	55.33
80	41.24	44.79	46.52	51.17	53.54	57.15	60.39	64.28
90	48.41	52.28	54.16	59.20	61.75	65.65	69.13	73.29
100	55.72	59.90	61.92	67.33	70.06	74.22	77.93	82.36
200	134.02	140.66	143.84	152.24	156.43	162.73	168.28	174.84
300	217.33	225.89	229.96	240.66	245.97	253.91	260.88	269.07
400	303.26	313.43	318.26	330.90	337.16	346.48	354.64	364.21
500	390.85	402.45	407.95	422.30	429.39	439.94	449.15	459.93
600	479.64	492.52	498.62	514.53	522.37	534.02	544.18	556.06
700	569.32	583.39	590.05	607.38	615.91	628.58	639.61	652.50
800	659.72	674.89	682.07	700.73	709.90	723.51	735.36	749.19
900	750.70	766.91	774.57	794.47	804.25	818.76	831.37	846.07
1000	842.17	859.36	867.48	888.56	898.91	914.26	927.59	943.13
1500	1304.80	1326.30	1336.42	1362.67	1375.53	1394.56	1411.06	1430.25
2000	1773.30	1798.42	1810.24	1840.85	1855.82	1877.95	1897.12	1919.39
2500	2245.54	2273.86	2287.17	2321.62	2338.45	2363.31	2384.84	2409.82
3000	2720.44	2751.65	2766.32	2804.23	2822.75	2850.08	2873.74	2901.17
3500	3197.36	3231.23	3247.14	3288.25	3308.31	3337.92	3363.53	3393.22
4000	3675.88	3712.22	3729.29	3773.37	3794.87	3826.60	3854.03	3885.81
4500	4155.71	4194.37	4212.52	4259.39	4282.25	4315.96	4345.10	4378.86
5000	4636.62	4677.48	4696.67	4746.17	4770.31	4805.90	4836.66	4872.28
5500	5118.47	5161.42	5181.58	5233.60	5258.96	5296.34	5328.63	5366.03
6000	5601.13	5646.08	5667.17	5721.59	5748.11	5787.20	5820.96	5860.05
6500	6084.50	6131.36	6153.35	6210.07	6237.70	6278.43	6313.60	6354.32
7000	6568.49	6617.20	6640.05	6698.98	6727.69	6769.99	6806.52	6848.80
7500	7053.05	7103.53	7127.22	7188.28	7218.03	7261.85	7299.69	7343.48
8000	7538.11	7590.32	7614.81	7677.94	7708.68	7753.98	7793.08	7838.33
8500	8023.63	8077.51	8102.78	8167.91	8199.63	8246.35	8286.68	8333.34
9000	8509.57	8565.07	8591.09	8658.17	8690.83	8738.94	8780.46	8828.50
9500	8995.90	9052.97	9079.73	9148.70	9182.28	9231.74	9274.42	9323.78
10000	9482.59	9541.19	9568.67	9639.48	9673.95	9724.72	9768.53	9819.19

6.2. CHI-SQUARE DISTRIBUTION

Critical values for the chi–square distribution $\chi^2_{\alpha,\nu}$.

ν	.10	.05	.025	α .01	.005	.001	.0005	.0001
1	2.7055	3.8415	5.0239	6.6349	7.8794	10.8276	12.1157	15.1367
2	4.6052	5.9915	7.3778	9.2103	10.5966	13.8155	15.2018	18.4207
3	6.2514	7.8147	9.3484	11.3449	12.8382	16.2662	17.7300	21.1075
4	7.7794	9.4877	11.1433	13.2767	14.8603	18.4668	19.9974	23.5127
5	9.2364	11.0705	12.8325	15.0863	16.7496	20.5150	22.1053	25.7448
6	10.6446	12.5916	14.4494	16.8119	18.5476	22.4577	24.1028	27.8563
7	12.0170	14.0671	16.0128	18.4753	20.2777	24.3219	26.0178	29.8775
8	13.3616	15.5073	17.5345	20.0902	21.9550	26.1245	27.8680	31.8276
9	14.6837	16.9190	19.0228	21.6660	23.5894	27.8772	29.6658	33.7199
10	15.9872	18.3070	20.4832	23.2093	25.1882	29.5883	31.4198	35.5640
11	17.2750	19.6751	21.9200	24.7250	26.7568	31.2641	33.1366	37.3670
12	18.5493	21.0261	23.3367	26.2170	28.2995	32.9095	34.8213	39.1344
13	19.8119	22.3620	24.7356	27.6882	29.8195	34.5282	36.4778	40.8707
14	21.0641	23.6848	26.1189	29.1412	31.3193	36.1233	38.1094	42.5793
15	22.3071	24.9958	27.4884	30.5779	32.8013	37.6973	39.7188	44.2632
16	23.5418	26.2962	28.8454	31.9999	34.2672	39.2524	41.3081	45.9249
17	24.7690	27.5871	30.1910	33.4087	35.7185	40.7902	42.8792	47.5664
18	25.9894	28.8693	31.5264	34.8053	37.1565	42.3124	44.4338	49.1894
19	27.2036	30.1435	32.8523	36.1909	38.5823	43.8202	45.9731	50.7955
20	28.4120	31.4104	34.1696	37.5662	39.9968	45.3147	47.4985	52.3860
21	29.6151	32.6706	35.4789	38.9322	41.4011	46.7970	49.0108	53.9620
22	30.8133	33.9244	36.7807	40.2894	42.7957	48.2679	50.5111	55.5246
23	32.0069	35.1725	38.0756	41.6384	44.1813	49.7282	52.0002	57.0746
24	33.1962	36.4150	39.3641	42.9798	45.5585	51.1786	53.4788	58.6130
25	34.3816	37.6525	40.6465	44.3141	46.9279	52.6197	54.9475	60.1403
26	35.5632	38.8851	41.9232	45.6417	48.2899	54.0520	56.4069	61.6573
27	36.7412	40.1133	43.1945	46.9629	49.6449	55.4760	57.8576	63.1645
28	37.9159	41.3371	44.4608	48.2782	50.9934	56.8923	59.3000	64.6624
29	39.0875	42.5570	45.7223	49.5879	52.3356	58.3012	60.7346	66.1517
30	40.2560	43.7730	46.9792	50.8922	53.6720	59.7031	62.1619	67.6326
31	41.4217	44.9853	48.2319	52.1914	55.0027	61.0983	63.5820	69.1057
32	42.5847	46.1943	49.4804	53.4858	56.3281	62.4872	64.9955	70.5712
33	43.7452	47.3999	50.7251	54.7755	57.6484	63.8701	66.4025	72.0296
34	44.9032	48.6024	51.9660	56.0609	58.9639	65.2472	67.8035	73.4812
35	46.0588	49.8018	53.2033	57.3421	60.2748	66.6188	69.1986	74.9262
36	47.2122	50.9985	54.4373	58.6192	61.5812	67.9852	70.5881	76.3650
37	48.3634	52.1923	55.6680	59.8925	62.8833	69.3465	71.9722	77.7977
38	49.5126	53.3835	56.8955	61.1621	64.1814	70.7029	73.3512	79.2247
39	50.6598	54.5722	58.1201	62.4281	65.4756	72.0547	74.7253	80.6462
40	51.8051	55.7585	59.3417	63.6907	66.7660	73.4020	76.0946	82.0623

Critical values for the chi–square distribution $\chi^2_{\alpha,\nu}$.

ν	.10	.05	.025	α .01	.005	.001	.0005	.0001
41	52.95	56.94	60.56	64.95	68.05	74.74	77.46	83.47
42	54.09	58.12	61.78	66.21	69.34	76.08	78.82	84.88
43	55.23	59.30	62.99	67.46	70.62	77.42	80.18	86.28
44	56.37	60.48	64.20	68.71	71.89	78.75	81.53	87.68
45	57.51	61.66	65.41	69.96	73.17	80.08	82.88	89.07
46	58.64	62.83	66.62	71.20	74.44	81.40	84.22	90.46
47	59.77	64.00	67.82	72.44	75.70	82.72	85.56	91.84
48	60.91	65.17	69.02	73.68	76.97	84.04	86.90	93.22
49	62.04	66.34	70.22	74.92	78.23	85.35	88.23	94.60
50	63.17	67.50	71.42	76.15	79.49	86.66	89.56	95.97
60	74.40	79.08	83.30	88.38	91.95	99.61	102.69	109.50
70	85.53	90.53	95.02	100.43	104.21	112.32	115.58	122.75
80	96.58	101.88	106.63	112.33	116.32	124.84	128.26	135.78
90	107.57	113.15	118.14	124.12	128.30	137.21	140.78	148.63
100	118.50	124.34	129.56	135.81	140.17	149.45	153.17	161.32
200	226.02	233.99	241.06	249.45	255.26	267.54	272.42	283.06
300	331.79	341.40	349.87	359.91	366.84	381.43	387.20	399.76
400	436.65	447.63	457.31	468.72	476.61	493.13	499.67	513.84
500	540.93	553.13	563.85	576.49	585.21	603.45	610.65	626.24
600	644.80	658.09	669.77	683.52	692.98	712.77	720.58	737.46
700	748.36	762.66	775.21	789.97	800.13	821.35	829.71	847.78
800	851.67	866.91	880.28	895.98	906.79	929.33	938.21	957.38
900	954.78	970.90	985.03	1001.63	1013.04	1036.83	1046.19	1066.40
1000	1057.72	1074.68	1089.53	1106.97	1118.95	1143.92	1153.74	1174.93
1500	1570.61	1591.21	1609.23	1630.35	1644.84	1674.97	1686.81	1712.30
2000	2081.47	2105.15	2125.84	2150.07	2166.66	2201.16	2214.68	2243.81
2500	2591.04	2617.43	2640.47	2667.43	2685.89	2724.22	2739.25	2771.57
3000	3099.69	3128.54	3153.70	3183.13	3203.28	3245.08	3261.45	3296.66
3500	3607.64	3638.75	3665.87	3697.57	3719.26	3764.26	3781.87	3819.74
4000	4115.05	4148.25	4177.19	4211.01	4234.14	4282.11	4300.88	4341.22
4500	4622.00	4657.17	4687.83	4723.63	4748.12	4798.87	4818.73	4861.40
5000	5128.58	5165.61	5197.88	5235.57	5261.34	5314.73	5335.62	5380.48
5500	5634.83	5673.64	5707.45	5746.93	5773.91	5829.81	5851.68	5898.63
6000	6140.81	6181.31	6216.59	6257.78	6285.92	6344.23	6367.02	6415.98
6500	6646.54	6688.67	6725.36	6768.18	6797.45	6858.05	6881.74	6932.61
7000	7152.06	7195.75	7233.79	7278.19	7308.53	7371.35	7395.90	7448.62
7500	7657.38	7702.58	7741.93	7787.86	7819.23	7884.18	7909.57	7964.06
8000	8162.53	8209.19	8249.81	8297.20	8329.58	8396.59	8422.78	8479.00
8500	8667.52	8715.59	8757.44	8806.26	8839.60	8908.62	8935.59	8993.48
9000	9172.36	9221.81	9264.85	9315.05	9349.34	9420.30	9448.03	9507.53
9500	9677.07	9727.86	9772.05	9823.60	9858.81	9931.67	9960.13	10021.21
10000	10181.66	10233.75	10279.07	10331.93	10368.03	10442.73	10471.91	10534.52

6.3 ERLANG DISTRIBUTION

6.3.1 Properties

pdf	$f(x) = \dfrac{x^{n-1} e^{-x/\beta}}{\beta^n (n-1)!},$	$x \geq 0,\ \beta > 0,\ n \in \mathcal{N}$
mean	$\mu = n\beta$	
variance	$\sigma^2 = n\beta^2$	
skewness	$\beta_1 = 2/\sqrt{n}$	
kurtosis	$\beta_2 = 3 + \dfrac{6}{n}$	
mgf	$m(t) = (1 - \beta t)^{-n}$	
char function	$\phi(t) = (1 - \beta i t)^{-n}$	

6.3.2 Probability density function

The probability density function is skewed to the right with n as the *shape* parameter.

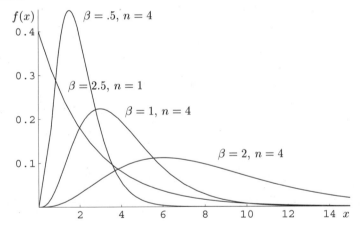

Figure 6.5: Probability density functions for an Erlang random variable.

6.3.3 Related distributions

If X is an Erlang random variable with parameters β and $n = 1$, then X is an exponential random variable with parameter $\lambda = 1/\beta$.

6.4 EXPONENTIAL DISTRIBUTION

6.4.1 Properties

pdf	$f(x) = \lambda e^{-\lambda x}$,	$x \geq 0$, $\lambda > 0$
mean	$\mu = 1/\lambda$	
variance	$\sigma^2 = 1/\lambda^2$	
skewness	$\beta_1 = 2$	
kurtosis	$\beta_2 = 9$	
mgf	$m(t) = \dfrac{\lambda}{\lambda - t}$	
char function	$\phi(t) = \dfrac{\lambda}{\lambda - it}$	

6.4.2 Probability density function

The probability density function is skewed to the right. The tail of the distribution is heavier for larger values of λ.

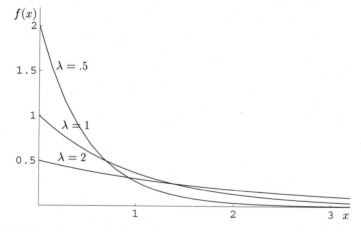

Figure 6.6: Probability density functions for an exponential random variable.

6.4.3 Related distributions

Let X be an exponential random variable with parameter λ.

(1) If $\lambda = 1/2$, then X is a chi-square random variable with $\nu = 2$.

(2) The random variable $\sqrt{X}$ has a Rayleigh distribution with parameter $\sigma = \sqrt{1/(2\lambda)}$.

(3) The random variable $Y = X^{1/\alpha}$ has a Weibull distribution with parameters α and $\lambda^{-1/\alpha}$.

6.5. F DISTRIBUTION

(4) The random variable $Y = e^{-X}$ has a power function distribution with parameters $b = 1$ and $c = \lambda$.

(5) The random variable $Y = ae^X$ has a Pareto distribution with parameters a and $\theta = \lambda$.

(6) The random variable $Y = \alpha - \ln X$ has an extreme-value distribution with parameters α and $\beta = 1/\lambda$.

(7) Let $X_1, X_2, \ldots, X_n$ be independent exponential random variables each with parameter λ.

   (a) The random variable $Y = \min(X_1, X_2, \ldots, X_n)$ has an exponential distribution with parameter $n\lambda$.

   (b) The random variable $Y = X_1 + X_2 + \cdots + X_n$ has an Erlang distribution with parameters $\beta = 1/\lambda$ and n.

(8) Let X_1 and X_2 be independent exponential random variables each with parameter λ. The random variable $Y = X_1 - X_2$ has a Laplace distribution with parameters 0 and $1/\lambda$.

(9) Let X be an exponential random variable with parameter $\lambda = 1$. The random variable $Y = -\ln[e^{-X}/(1 + e^{-X})]$ has a (standard) logistic distribution with parameters $\alpha = 0$ and $\beta = 1$.

(10) Let X_1 and X_2 be independent exponential random variables with parameter $\lambda = 1$.

   (a) The random variable $Y = X_1/(X_1 + X_2)$ has a (standard) uniform distribution with parameters $a = 0$ and $b = 1$.

   (b) The random variable $W = -\ln(X_1/X_2)$ has a (standard) logistic distribution with parameters $\alpha = 0$ and $\beta = 1$.

6.5 F DISTRIBUTION

6.5.1 Properties

$$\text{pdf} \quad f(x) = \frac{\Gamma\left(\frac{\nu_1+\nu_2}{2}\right) \nu_1^{\frac{\nu_1}{2}} \nu_2^{\frac{\nu_2}{2}}}{\Gamma(\nu_1/2)\Gamma(\nu_2/2)} x^{(\nu_1/2)-1} (\nu_2 + \nu_1 x)^{-(\nu_1+\nu_2)/2}$$

$$x > 0, \ \nu_1, \ \nu_2 > 0$$

$$\text{mean} \quad \mu = \frac{\nu_2}{\nu_2 - 2}, \quad \nu_2 \geq 3$$

$$\text{variance} \quad \sigma^2 = \frac{2\nu_2^2(\nu_1 + \nu_2 - 2)}{\nu_1(\nu_2 - 2)^2(\nu_2 - 4)}, \quad \nu_2 \geq 5$$

$$\text{skewness} \quad \beta_1 = \frac{(2\nu_1 + \nu_2 - 2)\sqrt{8(\nu_2 - 4)}}{\sqrt{\nu_1}(\nu_2 - 6)\sqrt{\nu_1 + \nu_2 - 2}}, \quad \nu_2 \geq 7$$

| kurtosis | $\beta_2 = 3 +$ |

$$\frac{12[(\nu_2-2)^2(\nu_2-4)+\nu_1(\nu_1+\nu_2-2)(5\nu_2-22)]}{\nu_1(\nu_2-6)(\nu_2-8)(\nu_1+\nu_2-2)}$$

$$\nu_2 \geq 9$$

| mgf | $m(t) =$ does not exist |
| char function | $\phi(t) = \Gamma\left(\dfrac{\nu_1+\nu_2}{2}\right)\Gamma\left(\dfrac{\nu_2}{2}\right)\psi\left(\dfrac{\nu_1}{2}, 1-\dfrac{\nu_2}{2}; \dfrac{it\nu_2}{\nu_1}\right)$ |

where $\Gamma(x)$ is the gamma function and ψ is the confluent hypergeometric function of the second kind.

6.5.2 Probability density function

The probability density function is skewed to the right with *shape* parameters ν_1 and ν_2. For fixed ν_2, the tail becomes lighter as ν_1 increases.

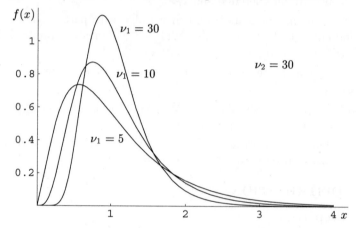

Figure 6.7: Probability density functions for an F random variable.

6.5.3 Related distributions

(1) If X has an F distribution with ν_1 and ν_2 degrees of freedom, then the random variable $Y = 1/X$ has an F distribution with ν_2 and ν_1 degrees of freedom.

(2) If X has an F distribution with ν_1 and ν_2 degrees of freedom, the random variable $\nu_1 X$ tends to a chi-square distribution with ν_1 degrees of freedom as $\nu_2 \to \infty$.

(3) Let X_1 and X_2 be independent F random variables with $\nu_1 = \nu_2 = \nu$ degrees of freedom. The random variable

$$Y = \frac{\sqrt{\nu}}{2}\left(\sqrt{X_1} - \sqrt{X_2}\right) \tag{6.6}$$

6.5. F DISTRIBUTION

has a t distribution with ν degrees of freedom.

(4) If X has an F distribution with parameters ν_1 and ν_2, the random variable

$$Y = \frac{\nu_1 X/\nu_2}{1 + \frac{\nu_1 X}{\nu_2}} \qquad (6.7)$$

has a beta distribution with parameters $\alpha = \nu_2/2$ and $\beta = \nu_1/2$.

6.5.4 Critical values for the F distribution

Given values of ν_1, ν_2, and α, the tables on pages 92–95 contain values of F_{α,ν_1,ν_2} such that

$$\begin{aligned} 1 - \alpha &= \int_0^{F_{\alpha,\nu_1,\nu_2}} f(x)\,dx \\ &= \int_0^{F_{\alpha,\nu_1,\nu_2}} \frac{\Gamma\left(\frac{\nu_1+\nu_2}{2}\right) \nu_1^{\frac{\nu_1}{2}} \nu_2^{\frac{\nu_2}{2}}}{\Gamma(\nu_1/2)\Gamma(\nu_2/2)} x^{(\nu_1/2)-1} (\nu_2 + \nu_1 x)^{-(\nu_1+\nu_2)/2}\,dx \end{aligned} \qquad (6.8)$$

Note that $F_{1-\alpha}$ for ν_1 and ν_2 degrees of freedom is the reciprocal of F_α for ν_2 and ν_1 degrees of freedom. For example,

$$F_{.05,4,7} = \frac{1}{F_{.95,7,4}} = \frac{1}{6.09} = .164 \qquad (6.9)$$

Example 6.38: Use the following tables to find the values $F_{.1,4,9}$ and $F_{.95,12,15}$.

Solution:

(S1) The top rows of the tables on pages 92–95 contain entries for the numerator degrees of freedom and the left-hand column contains the denominator degrees of freedom. The intersection of the ν_1 degrees of freedom column and the ν_2 row may be used to find critical values of the form F_{α,ν_1,ν_2} such that $\text{Prob}\,[F \geq F_{\alpha,\nu_1,\nu_2}] = \alpha$.

(S2) $F_{.1,4,9} = 2.69 \implies \text{Prob}\,[F \geq 2.69] = .1$

$F_{.95,12,15} = \dfrac{1}{F_{.05,15,12}} = \dfrac{1}{2.62} = .3817 \implies \text{Prob}\,[F \geq .3817] = .95$

(S3) Illustrations:

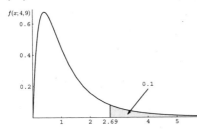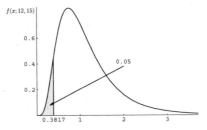

Critical values for the F distribution

For given values of ν_1 and ν_2, the following table contains values of $F_{0.1,\nu_1,\nu_2}$; defined by $\text{Prob}\left[F \geq F_{0.1,\nu_1,\nu_2}\right] = \alpha = 0.1$.

ν_2	$\nu_1=1$	2	3	4	5	6	7	8	9	10	50	100	∞
1	39.86	49.50	53.59	55.83	57.24	58.20	58.91	59.44	59.86	60.19	62.69	63.01	63.33
2	8.53	9.00	9.16	9.24	9.29	9.33	9.35	9.37	9.38	9.39	9.47	9.48	9.49
3	5.54	5.46	5.39	5.34	5.31	5.28	5.27	5.25	5.24	5.23	5.15	5.14	5.13
4	4.54	4.32	4.19	4.11	4.05	4.01	3.98	3.95	3.94	3.92	3.80	3.78	3.76
5	4.06	3.78	3.62	3.52	3.45	3.40	3.37	3.34	3.32	3.30	3.15	3.13	3.10
6	3.78	3.46	3.29	3.18	3.11	3.05	3.01	2.98	2.96	2.94	2.77	2.75	2.72
7	3.59	3.26	3.07	2.96	2.88	2.83	2.78	2.75	2.72	2.70	2.52	2.50	2.47
8	3.46	3.11	2.92	2.81	2.73	2.67	2.62	2.59	2.56	2.54	2.35	2.32	2.29
9	3.36	3.01	2.81	2.69	2.61	2.55	2.51	2.47	2.44	2.42	2.22	2.19	2.16
10	3.29	2.92	2.73	2.61	2.52	2.46	2.41	2.38	2.35	2.32	2.12	2.09	2.06
11	3.23	2.86	2.66	2.54	2.45	2.39	2.34	2.30	2.27	2.25	2.04	2.01	1.97
12	3.18	2.81	2.61	2.48	2.39	2.33	2.28	2.24	2.21	2.19	1.97	1.94	1.90
13	3.14	2.76	2.56	2.43	2.35	2.28	2.23	2.20	2.16	2.14	1.92	1.88	1.85
14	3.10	2.73	2.52	2.39	2.31	2.24	2.19	2.15	2.12	2.10	1.87	1.83	1.80
15	3.07	2.70	2.49	2.36	2.27	2.21	2.16	2.12	2.09	2.06	1.83	1.79	1.76
16	3.05	2.67	2.46	2.33	2.24	2.18	2.13	2.09	2.06	2.03	1.79	1.76	1.72
17	3.03	2.64	2.44	2.31	2.22	2.15	2.10	2.06	2.03	2.00	1.76	1.73	1.69
18	3.01	2.62	2.42	2.29	2.20	2.13	2.08	2.04	2.00	1.98	1.74	1.70	1.66
19	2.99	2.61	2.40	2.27	2.18	2.11	2.06	2.02	1.98	1.96	1.71	1.67	1.63
20	2.97	2.59	2.38	2.25	2.16	2.09	2.04	2.00	1.96	1.94	1.69	1.65	1.61
25	2.92	2.53	2.32	2.18	2.09	2.02	1.97	1.93	1.89	1.87	1.61	1.56	1.52
50	2.81	2.41	2.20	2.06	1.97	1.90	1.84	1.80	1.76	1.73	1.44	1.39	1.34
100	2.76	2.36	2.14	2.00	1.91	1.83	1.78	1.73	1.69	1.66	1.35	1.29	1.20
∞	2.71	2.30	2.08	1.94	1.85	1.77	1.72	1.67	1.63	1.60	1.24	1.17	1.00

6.5. F DISTRIBUTION

Critical values for the F distribution

For given values of ν_1 and ν_2, the following table contains values of $F_{0.05,\nu_1,\nu_2}$; defined by $\text{Prob}[F \geq F_{0.05,\nu_1,\nu_2}] = \alpha = 0.05$.

ν_2	$\nu_1=1$	2	3	4	5	6	7	8	9	10	50	100	∞
1	161.4	199.5	215.7	224.6	230.2	234.0	236.8	238.9	240.5	241.9	251.8	253.0	254.3
2	18.51	19.00	19.16	19.25	19.30	19.33	19.35	19.37	19.38	19.40	19.48	19.49	19.50
3	10.13	9.55	9.28	9.12	9.01	8.94	8.89	8.85	8.81	8.79	8.58	8.55	8.53
4	7.71	6.94	6.59	6.39	6.26	6.16	6.09	6.04	6.00	5.96	5.70	5.66	5.63
5	6.61	5.79	5.41	5.19	5.05	4.95	4.88	4.82	4.77	4.74	4.44	4.41	4.36
6	5.99	5.14	4.76	4.53	4.39	4.28	4.21	4.15	4.10	4.06	3.75	3.71	3.67
7	5.59	4.74	4.35	4.12	3.97	3.87	3.79	3.73	3.68	3.64	3.32	3.27	3.23
8	5.32	4.46	4.07	3.84	3.69	3.58	3.50	3.44	3.39	3.35	3.02	2.97	2.93
9	5.12	4.26	3.86	3.63	3.48	3.37	3.29	3.23	3.18	3.14	2.80	2.76	2.71
10	4.96	4.10	3.71	3.48	3.33	3.22	3.14	3.07	3.02	2.98	2.64	2.59	2.54
11	4.84	3.98	3.59	3.36	3.20	3.09	3.01	2.95	2.90	2.85	2.51	2.46	2.40
12	4.75	3.89	3.49	3.26	3.11	3.00	2.91	2.85	2.80	2.75	2.40	2.35	2.30
13	4.67	3.81	3.41	3.18	3.03	2.92	2.83	2.77	2.71	2.67	2.31	2.26	2.21
14	4.60	3.74	3.34	3.11	2.96	2.85	2.76	2.70	2.65	2.60	2.24	2.19	2.13
15	4.54	3.68	3.29	3.06	2.90	2.79	2.71	2.64	2.59	2.54	2.18	2.12	2.07
16	4.49	3.63	3.24	3.01	2.85	2.74	2.66	2.59	2.54	2.49	2.12	2.07	2.01
17	4.45	3.59	3.20	2.96	2.81	2.70	2.61	2.55	2.49	2.45	2.08	2.02	1.96
18	4.41	3.55	3.16	2.93	2.77	2.66	2.58	2.51	2.46	2.41	2.04	1.98	1.92
19	4.38	3.52	3.13	2.90	2.74	2.63	2.54	2.48	2.42	2.38	2.00	1.94	1.88
20	4.35	3.49	3.10	2.87	2.71	2.60	2.51	2.45	2.39	2.35	1.97	1.91	1.84
25	4.24	3.39	2.99	2.76	2.60	2.49	2.40	2.34	2.28	2.24	1.84	1.78	1.71
50	4.03	3.18	2.79	2.56	2.40	2.29	2.20	2.13	2.07	2.03	1.60	1.52	1.45
100	3.94	3.09	2.70	2.46	2.31	2.19	2.10	2.03	1.97	1.93	1.48	1.39	1.28
∞	3.84	3.00	2.60	2.37	2.21	2.10	2.01	1.94	1.88	1.83	1.35	1.25	1.00

Critical values for the F distribution

For given values of ν_1 and ν_2, the following table contains values of $F_{0.01,\nu_1,\nu_2}$; defined by $\text{Prob}\,[F \geq F_{0.01,\nu_1,\nu_2}] = \alpha = 0.01$.

ν_2	$\nu_1=1$	2	3	4	5	6	7	8	9	10	50	100	∞
1	4052	5000	5403	5625	5764	5859	5928	5981	6022	6056	6303	6334	6336
2	98.50	99.00	99.17	99.25	99.30	99.33	99.36	99.37	99.39	99.40	99.48	99.49	99.50
3	34.12	30.82	29.46	28.71	28.24	27.91	27.67	27.49	27.35	27.23	26.35	26.24	26.13
4	21.20	18.00	16.69	15.98	15.52	15.21	14.98	14.80	14.66	14.55	13.69	13.58	13.46
5	16.26	13.27	12.06	11.39	10.97	10.67	10.46	10.29	10.16	10.05	9.24	9.13	9.02
6	13.75	10.92	9.78	9.15	8.75	8.47	8.26	8.10	7.98	7.87	7.09	6.99	6.88
7	12.25	9.55	8.45	7.85	7.46	7.19	6.99	6.84	6.72	6.62	5.86	5.75	5.65
8	11.26	8.65	7.59	7.01	6.63	6.37	6.18	6.03	5.91	5.81	5.07	4.96	4.86
9	10.56	8.02	6.99	6.42	6.06	5.80	5.61	5.47	5.35	5.26	4.52	4.41	4.31
10	10.04	7.56	6.55	5.99	5.64	5.39	5.20	5.06	4.94	4.85	4.12	4.01	3.91
11	9.65	7.21	6.22	5.67	5.32	5.07	4.89	4.74	4.63	4.54	3.81	3.71	3.60
12	9.33	6.93	5.95	5.41	5.06	4.82	4.64	4.50	4.39	4.30	3.57	3.47	3.36
13	9.07	6.70	5.74	5.21	4.86	4.62	4.44	4.30	4.19	4.10	3.38	3.27	3.17
14	8.86	6.51	5.56	5.04	4.69	4.46	4.28	4.14	4.03	3.94	3.22	3.11	3.00
15	8.68	6.36	5.42	4.89	4.56	4.32	4.14	4.00	3.89	3.80	3.08	2.98	2.87
16	8.53	6.23	5.29	4.77	4.44	4.20	4.03	3.89	3.78	3.69	2.97	2.86	2.75
17	8.40	6.11	5.18	4.67	4.34	4.10	3.93	3.79	3.68	3.59	2.87	2.76	2.65
18	8.29	6.01	5.09	4.58	4.25	4.01	3.84	3.71	3.60	3.51	2.78	2.68	2.57
19	8.18	5.93	5.01	4.50	4.17	3.94	3.77	3.63	3.52	3.43	2.71	2.60	2.49
20	8.10	5.85	4.94	4.43	4.10	3.87	3.70	3.56	3.46	3.37	2.64	2.54	2.42
25	7.77	5.57	4.68	4.18	3.85	3.63	3.46	3.32	3.22	3.13	2.40	2.29	2.17
50	7.17	5.06	4.20	3.72	3.41	3.19	3.02	2.89	2.78	2.70	1.95	1.82	1.70
100	6.90	4.82	3.98	3.51	3.21	2.99	2.82	2.69	2.59	2.50	1.74	1.60	1.45
∞	6.63	4.61	3.78	3.32	3.02	2.80	2.64	2.51	2.41	2.32	1.53	1.32	1.00

Critical values for the F distribution

For given values of ν_1 and ν_2, the following table contains values of $F_{0.001,\nu_1,\nu_2}$; defined by $\text{Prob}\left[F \geq F_{0.001,\nu_1,\nu_2}\right] = \alpha = 0.001$.

ν_2	$\nu_1 = 1$	2	3	4	5	6	7	8	9	10	50	100	∞
2	998.5	999.0	999.2	999.2	999.3	999.3	999.4	999.4	999.4	999.4	999.5	999.5	999.5
3	167.0	148.5	141.1	137.1	134.6	132.8	131.6	130.6	129.9	129.2	124.7	124.1	123.5
4	74.14	61.25	56.18	53.44	51.71	50.53	49.66	49.00	48.47	48.05	44.88	44.47	44.05
5	47.18	37.12	33.20	31.09	29.75	28.83	28.16	27.65	27.24	26.92	24.44	24.12	23.79
6	35.51	27.00	23.70	21.92	20.80	20.03	19.46	19.03	18.69	18.41	16.31	16.03	15.75
7	29.25	21.69	18.77	17.20	16.21	15.52	15.02	14.63	14.33	14.08	12.20	11.95	11.70
8	25.41	18.49	15.83	14.39	13.48	12.86	12.40	12.05	11.77	11.54	9.80	9.57	9.33
9	22.86	16.39	13.90	12.56	11.71	11.13	10.70	10.37	10.11	9.89	8.26	8.04	7.81
10	21.04	14.91	12.55	11.28	10.48	9.93	9.52	9.20	8.96	8.75	7.19	6.98	6.76
11	19.69	13.81	11.56	10.35	9.58	9.05	8.66	8.35	8.12	7.92	6.42	6.21	6.00
12	18.64	12.97	10.80	9.63	8.89	8.38	8.00	7.71	7.48	7.29	5.83	5.63	5.42
13	17.82	12.31	10.21	9.07	8.35	7.86	7.49	7.21	6.98	6.80	5.37	5.17	4.97
14	17.14	11.78	9.73	8.62	7.92	7.44	7.08	6.80	6.58	6.40	5.00	4.81	4.60
15	16.59	11.34	9.34	8.25	7.57	7.09	6.74	6.47	6.26	6.08	4.70	4.51	4.31
16	16.12	10.97	9.01	7.94	7.27	6.80	6.46	6.19	5.98	5.81	4.45	4.26	4.06
17	15.72	10.66	8.73	7.68	7.02	6.56	6.22	5.96	5.75	5.58	4.24	4.05	3.85
18	15.38	10.39	8.49	7.46	6.81	6.35	6.02	5.76	5.56	5.39	4.06	3.87	3.67
19	15.08	10.16	8.28	7.27	6.62	6.18	5.85	5.59	5.39	5.22	3.90	3.71	3.51
20	14.82	9.95	8.10	7.10	6.46	6.02	5.69	5.44	5.24	5.08	3.77	3.58	3.38
25	13.88	9.22	7.45	6.49	5.89	5.46	5.15	4.91	4.71	4.56	3.28	3.09	2.89
50	12.22	7.96	6.34	5.46	4.90	4.51	4.22	4.00	3.82	3.67	2.44	2.25	2.06
100	11.50	7.41	5.86	5.02	4.48	4.11	3.83	3.61	3.44	3.30	2.08	1.87	1.65
∞	10.83	6.91	5.42	4.62	4.10	3.74	3.47	3.27	3.10	2.96	1.75	1.45	1.00

6.6 GAMMA DISTRIBUTION

6.6.1 Properties

pdf	$f(x) = \dfrac{x^{\alpha-1} e^{-x/\beta}}{\beta^\alpha \Gamma(\alpha)},$	$x \geq 0,\ \alpha > 0,\ \beta > 0$
mean	$\mu = \alpha\beta$	
variance	$\sigma^2 = \alpha\beta^2$	
skewness	$\beta_1 = 2/\sqrt{\alpha}$	
kurtosis	$\beta_2 = 3\left(1 + \dfrac{2}{\alpha}\right)$	
mgf	$m(t) = (1 - \beta t)^{-\alpha}$	
char function	$\phi(t) = (1 - i\beta t)^{-\alpha}$	

where $\Gamma(x)$ is the gamma function (see page 204).

6.6.2 Probability density function

The probability density function is skewed to the right. For fixed β the tail becomes heavier as α increases.

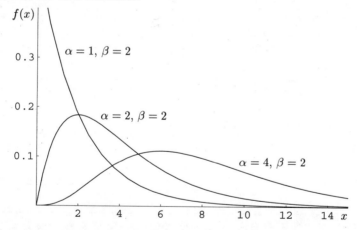

Figure 6.8: Probability density functions for a gamma random variable.

6.6.3 Related distributions

Let X be a gamma random variable with parameters α and β.

(1) The random variable X has a standard gamma distribution if $\alpha = 1$.

(2) If $\alpha = 1$ and $\beta = 1/\lambda$, then X has an exponential distribution with parameter λ.

(3) If $\alpha = \nu/2$ and $\beta = 2$, then X has a chi-square distribution with ν degrees of freedom.

(4) If $\alpha = n$ is an integer, then X has an Erlang distribution with parameters β and n.

(5) If $\alpha = \nu/2$ and $\beta = 1$, then the random variable $Y = 2X$ has a chi-square distribution with ν degrees of freedom.

(6) As $\alpha \to \infty$, X tends to a normal distribution with parameters $\mu = \alpha\beta$ and $\sigma^2 = \alpha\beta^2$.

(7) Suppose X_1 is a gamma random variable with parameters $\alpha = 1$ and $\beta = \beta_1$, X_2 is a gamma random variable with parameters $\alpha = 1$ and $\beta = \beta_2$, and X_1 and X_2 are independent. The random variable $Y = X_1/(X_1 + X_2)$ has a beta distribution with parameters β_1 and β_2.

(8) Let $X_1, X_2, \ldots, X_n$ be independent gamma random variables with parameters α_i and β for $i = 1, 2, \ldots, n$. The random variable $Y = X_1 + X_2 + \cdots + X_n$ has a gamma distribution with parameters $\alpha = \alpha_1 + \alpha_2 + \cdots + \alpha_n$ and β.

6.7 LOGNORMAL DISTRIBUTION

6.7.1 Properties

$$\text{pdf} \quad f(x) = \frac{1}{\sqrt{2\pi}\,\sigma x} \exp\left(-\frac{1}{2\sigma^2}(\ln x - \mu)^2\right)$$

$$x > 0,\ \mu \in \mathcal{R},\ \sigma > 0$$

mean $\quad \mu = e^{\mu + \sigma^2/2}$

variance $\quad \sigma^2 = e^{2\mu + \sigma^2}(e^{\sigma^2} - 1)$

skewness $\quad \beta_1 = (e^{\sigma^2} + 2)\sqrt{e^{\sigma^2} - 1}$

kurtosis $\quad \beta_2 = e^{4\sigma^2} + 2e^{3\sigma^2} + 3e^{2\sigma^2}$

mgf $\quad m(t) = $ does not exist

char function $\quad \phi(t) = $ does not exist

6.7.2 Probability density function

The probability density function is skewed to the right. The *scale* parameter is μ and the *shape* parameter is σ.

6.7.3 Related distributions

(1) If X is a lognormal random variable with parameters μ and σ, then the random variable $Y = \ln X$ has a normal distribution with mean μ and variance σ^2.

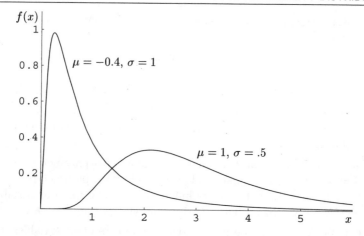

Figure 6.9: Probability density functions for a lognormal random variable.

(2) If X is a lognormal random variable with parameters μ and σ and a and b are constants, then the random variable $Y = e^a X^b$ has a lognormal distribution with parameters $a + b\mu$ and $b\sigma$.

(3) Let X_1 and X_2 be independent lognormal random variables with parameters μ_1, σ_1 and μ_2, σ_2, respectively. The random variable $Y = X_1/X_2$ has a lognormal distribution with parameters $\mu_1 - \mu_2$ and $\sigma_1 + \sigma_2$.

(4) Let $X_1, X_2, \ldots, X_n$ be independent lognormal random variables with parameters μ_i and σ_i for $i = 1, 2, \ldots, n$. The random variable $Y = X_1 \cdot X_2 \cdots X_n$ has a lognormal distribution with parameters $\mu = \mu_1 + \mu_2 + \cdots + \mu_n$ and $\sigma = \sigma_1 + \sigma_2 + \cdots + \sigma_n$.

(5) Let $X_1, X_2, \ldots, X_n$ be independent lognormal random variables with parameters μ and σ. The random variable $Y = \sqrt[n]{X_1 \cdots X_n}$ has a lognormal distribution with parameters μ and σ/n.

6.8 NORMAL DISTRIBUTION

6.8.1 Properties

$$\text{pdf} \quad f(x) = \frac{1}{\sigma\sqrt{2\pi}} e^{-(x-\mu)^2/2\sigma^2}, \quad x \in \mathcal{R},\ \mu \in \mathcal{R},\ \sigma > 0$$

mean $\quad \mu = \mu$

variance $\quad \sigma^2 = \sigma^2$

skewness $\quad \beta_1 = 0$

kurtosis $\quad \beta_2 = 3$

$$\text{mgf} \quad m(t) = \exp\left(\mu t + \frac{\sigma^2 t^2}{2}\right)$$

$$\text{char function} \quad \phi(t) = \exp\left(\mu i t - \frac{\sigma^2 t^2}{2}\right)$$

See Chapter 7 for more details.

6.8.2 Probability density function

The probability density function is symmetric and bell-shaped about the *location* parameter μ. For small values of the *scale* parameter σ the probability density function is more compact.

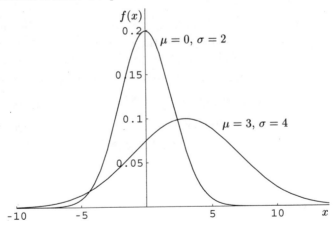

Figure 6.10: Probability density functions for a normal random variable.

6.8.3 Related distributions

(1) The random variable X has a standard normal distribution if $\mu = 0$ and $\sigma = 1$.

(2) If X is a normal random variable with parameters μ and σ, the random variable $Y = (X - \mu)/\sigma$ has a (standard) normal distribution with parameters 0 and 1.

(3) If X is a normal random variable with parameters μ and σ, the random variable $Y = e^X$ has a lognormal distribution with parameters μ and σ.

(4) If X is a normal random variable with parameters $\mu = 0$ and $\sigma = 1$, then the random variable $Y = e^{\mu + \sigma X}$ has a lognormal distribution with parameters μ and σ.

(5) If X is a normal random variable with parameters μ and σ, and a and b are constants, then the random variable $Y = a + bX$ has a normal distribution with parameters $a + b\mu$ and $b\sigma$.

(6) If X_1 and X_2 are independent standard normal random variables, the random variable $Y = X_1/X_2$ has a Cauchy distribution with parameters $a = 0$ and $b = 1$.

(7) If X_1 and X_2 are independent normal random variables with parameters $\mu = 0$ and σ, then the random variable $Y = \sqrt{X_1^2 + X_2^2}$ has a Rayleigh distribution with parameter σ.

(8) Let X_i (for $i = 1, 2, \ldots, n$) be independent, normal random variables with parameters μ_i and σ_i, and let c_i be any constants. The random variable $Y = \sum_{i=1}^{n} c_i X_i$ has a normal distribution with parameters $\mu = \sum_{i=1}^{n} c_i \mu_i$ and $\sigma^2 = \sum_{i=1}^{n} c_i^2 \sigma_i^2$.

(9) Let X_i (for $i = 1, 2, \ldots, n$) be independent, normal random variables with parameters μ and σ, then the random variable $Y = X_1 + X_2 + \cdots + X_n$ has a normal distribution with mean $n\mu$ and variance $n\sigma^2$.

(10) Let X_i (for $i = 1, 2, \ldots, n$) be independent standard normal random variables. The random variable $Y = \sum_{i=1}^{n} X_i^2$ has a chi-square distribution with $\nu = n$ degrees of freedom. If $\mu_i = \lambda_i > 0$ ($\sigma_i = 1$), then the random variable Y has a noncentral chi-square distribution with parameters $\nu = n$ and noncentrality parameter $\lambda = \sum_{i=1}^{n} \lambda_i^2$.

6.9 NORMAL DISTRIBUTION: MULTIVARIATE

6.9.1 Properties

$$\text{pdf} \quad f(\mathbf{x}) = \frac{1}{(2\pi)^{n/2}\sqrt{\det(\Sigma)}} \exp\left[-\frac{(\mathbf{x}-\boldsymbol{\mu})^{\mathrm{T}}\Sigma^{-1}(\mathbf{x}-\boldsymbol{\mu})}{2}\right]$$

$$\text{mean} \quad \boldsymbol{\mu}$$

$$\text{covariance matrix} \quad \Sigma$$

$$\text{char function} \quad \phi(\mathbf{t}) = \exp\left(i\mathbf{t}^{\mathrm{T}}\boldsymbol{\mu} - \frac{1}{2}\mathbf{t}^{\mathrm{T}}\Sigma\mathbf{t}\right)$$

where $\mathbf{x} = [x_1, x_2, \ldots, x_n]^{\mathrm{T}}$ (with $x_i \in \mathcal{R}$) and Σ is a positive semi-definite matrix.

Section 7.6 discusses the bivariate normal.

6.9.2 Probability density function

The probability density function is smooth and unimodal. Figure 6.11 shows two views of a bivariate normal with $\boldsymbol{\mu} = \begin{bmatrix} 1 & 0 \end{bmatrix}^{\mathrm{T}}$ and $\Sigma = \begin{bmatrix} 1 & 2 \\ 0 & 4 \end{bmatrix}$.

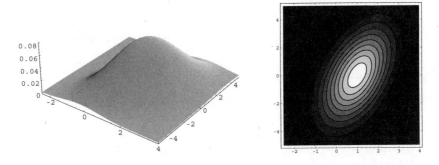

Figure 6.11: Two views of the probability density for a bivariate normal.

6.10 PARETO DISTRIBUTION

6.10.1 Properties

pdf $f(x) = \dfrac{\theta a^\theta}{x^{\theta+1}}, \quad x \geq a, \; \theta > 0, \; a > 0$

mean $\mu = \dfrac{a\theta}{\theta - 1}, \quad \theta > 1$

variance $\sigma^2 = \dfrac{a^2\theta}{(\theta-1)^2(\theta-2)}, \quad \theta > 2$

skewness $\beta_1 = \dfrac{2(\theta+1)\sqrt{\theta-2}}{(\theta-3)\sqrt{\theta}}, \quad \theta > 3$

kurtosis $\beta_2 = \dfrac{3(\theta-2)(3\theta^2+\theta+2)}{\theta(\theta-3)(\theta-4)}, \quad \theta > 4$

mgf $m(t) = $ does not exist

char function $\phi(t) = -a^\theta t^\theta \cos(\pi\theta/2)\Gamma(1-\theta) +$
$\quad {}_1F_2\left[\left\{-\frac{\theta}{2}\right\},\left\{\frac{1}{2}, 1-\frac{\theta}{2}\right\}, -\frac{1}{4}a^2 t^2\right] -$
$\quad \frac{1}{1-\theta}\left(ati\theta_1 F_2\left[\left\{\frac{1}{2}-\frac{\theta}{2}\right\},\left\{\frac{3}{2},\frac{3}{2}-\frac{\theta}{2}\right\}, -\frac{1}{4}a^2 t^2\right]\operatorname{sgn}(t)\right) +$
$\quad ia^\theta t^\theta \Gamma(1-\theta)\operatorname{sgn}(t)\sin(\pi\theta/2)$

where ${}_pF_q$ is the generalized hypergeometric function and $\operatorname{sgn}(t)$ is the signum function.

6.10.2 Probability density function

The probability density function is skewed to the right. The *shape* parameter is θ and the *location* parameter is a.

6.10.3 Related distributions

(1) Let X be a Pareto random variable with parameters a and θ.
   (a) The random variable $Y = \ln(X/a)$ has an exponential distribution with parameter $\lambda = 1/\theta$.
   (b) The random variable $Y = 1/X$ has a power function distribution with parameters $1/a$ and θ.
   (c) The random variable $Y = -\ln\left[(X/a)^\theta - 1\right]$ has a logistic distribution with parameters $\alpha = 0$ and $\beta = 1$.

(2) Let X_i (for $i = 1, 2, \ldots, n$) be independent Pareto random variables with parameters a and θ. The random variable $Y = 2a\sum_{i=1}^{n}\ln(X_i/\theta)$ has a chi-square distribution with $\nu = 2n$.

6.11. RAYLEIGH DISTRIBUTION

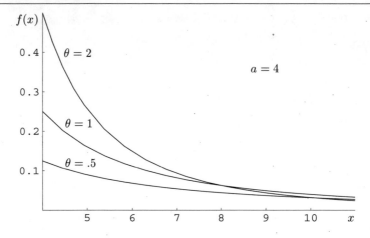

Figure 6.12: Probability density functions for a Pareto random variable.

6.11 RAYLEIGH DISTRIBUTION

6.11.1 Properties

pdf	$f(x) = \dfrac{x}{\sigma^2} \exp\left(-\dfrac{x^2}{2\sigma^2}\right),$	$x \geq 0,\ \sigma > 0$
mean	$\mu = \sigma\sqrt{\pi/2}$	
variance	$\sigma^2 = \sigma^2\left(2 - \dfrac{\pi}{2}\right)$	
skewness	$\beta_1 = \dfrac{(\pi-3)\sqrt{\pi/2}}{\left(2-\frac{\pi}{2}\right)^{3/2}}$	
kurtosis	$\beta_2 = \dfrac{32 - 3\pi^2}{(4-\pi)^2}$	
mgf	$m(t) = \dfrac{1}{2}\left(2 + \sqrt{2\pi}\,\sigma\,t\,e^{\sigma^2 t^2/2}\left[1 + \operatorname{erf}\left(\dfrac{\sigma t}{\sqrt{2}}\right)\right]\right)$	
char function	$\phi(t) = 1 + i e^{-\sigma^2 t^2/2}\sqrt{\dfrac{\pi}{2}}\,\sigma t\left[1 - \operatorname{erf}\left(-\dfrac{i\sigma t}{\sqrt{2}}\right)\right]$	

where $\operatorname{erf}(x)$ is the error function (see page 201).

6.11.2 Probability density function

The probability density function is skewed to the right. For large values of σ the tail is heavier.

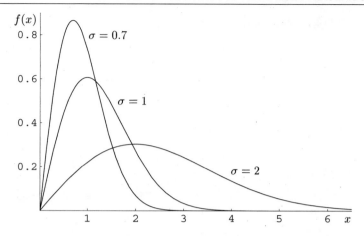

Figure 6.13: Probability density functions for a Rayleigh random variable.

6.11.3 Related distributions

(1) If X is a Rayleigh random variable with parameter $\sigma = 1$, then X is a chi random variable with parameter $n = 2$.

(2) If X is a Rayleigh random variable with parameter σ, then the random variable $Y = X^2$ has an exponential distribution with parameter $\lambda = 1/(2\sigma^2)$.

6.12 t DISTRIBUTION

6.12.1 Properties

$$\text{pdf} \quad f(x) = \frac{1}{\sqrt{\pi\nu}} \frac{\Gamma\left(\frac{\nu+1}{2}\right)}{\Gamma\left(\frac{\nu}{2}\right)} \left(1 + \frac{x^2}{\nu}\right)^{-(\nu+1)/2} \quad x \in \mathcal{R},\ \nu \in \mathcal{N}$$

$$\text{mean} \quad \mu = 0,\quad \nu \geq 2$$

$$\text{variance} \quad \sigma^2 = \frac{\nu}{\nu - 2},\quad \nu \geq 3$$

$$\text{skewness} \quad \beta_1 = 0,\quad \nu \geq 4$$

$$\text{kurtosis} \quad \beta_2 = 3 + \frac{6}{\nu - 4},\quad \nu \geq 5$$

$$\text{mgf} \quad m(t) = \text{does not exist}$$

$$\text{char function} \quad \phi(t) = \frac{2^{1-\frac{\nu}{2}} \nu^{\nu/4} |t|^{\nu/2} K_{\nu/2}(\sqrt{\nu}|t|)}{\Gamma(\nu/2)}$$

where $K_n(x)$ is a modified Bessel function and $\Gamma(x)$ is the gamma function.

6.12.2 Probability density function

The probability density function is symmetric and bell-shaped centered about 0. As the degrees of freedom, ν, increases the distribution becomes more compact.

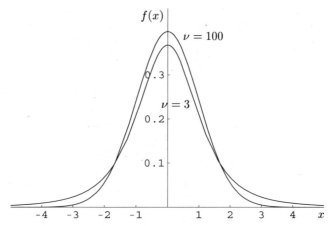

Figure 6.14: Probability density functions for a t random variable.

6.12.3 Related distributions

(1) If X is a t random variable with parameter ν, then the random variable $Y = X^2$ has an F distribution with 1 and ν degrees of freedom.

(2) If X is a t random variable with parameter $\nu = 1$, then X has a Cauchy distribution with parameters $a = 0$ and $b = 1$.

(3) If X is a t random variable with parameter ν, as ν tends to infinity X tends to a standard normal distribution. The approximation is reasonable for $\nu \geq 30$.

6.12.4 Critical values for the t distribution

For a given value of ν, the number of degrees of freedom, the table on page 106 contains values of $t_{\alpha,\nu}$ such that

$$\text{Prob}\left[t \geq t_{\alpha,\nu}\right] = \alpha \qquad (6.10)$$

Example 6.39: Use the table on page 106 to find the values $t_{.05,11}$ and $-t_{.01,24}$.
Solution:

(S1) The top row of the following table contains cumulative probability and the left-hand column contains the degrees of freedom. The values in the body of the table may be used to find critical values.

(S2) $t_{.05,11} = 1.7959$ since $F(1.7959; 11) = .95 \implies \text{Prob}\left[t \geq 1.7959\right] = .05$

(S3) $-t_{.01,24} = -2.4922$ since $F(2.4922; 24) = .99$
$\implies \text{Prob}\left[t \leq -2.4922\right] = .01$

Critical values for the t distribution.

ν	$\alpha = 0.1$	0.05	0.025	0.01	0.005	0.0025	0.001
1	3.078	6.314	12.706	31.821	63.657	318.309	636.619
2	1.886	2.920	4.303	6.965	9.925	22.327	31.599
3	1.638	2.353	3.182	4.541	5.841	10.215	12.924
4	1.533	2.132	2.776	3.747	4.604	7.173	8.610
5	1.476	2.015	2.571	3.365	4.032	5.893	6.869
6	1.440	1.943	2.447	3.143	3.707	5.208	5.959
7	1.415	1.895	2.365	2.998	3.499	4.785	5.408
8	1.397	1.860	2.306	2.896	3.355	4.501	5.041
9	1.383	1.833	2.262	2.821	3.250	4.297	4.781
10	1.372	1.812	2.228	2.764	3.169	4.144	4.587
11	1.363	1.796	2.201	2.718	3.106	4.025	4.437
12	1.356	1.782	2.179	2.681	3.055	3.930	4.318
13	1.350	1.771	2.160	2.650	3.012	3.852	4.221
14	1.345	1.761	2.145	2.624	2.977	3.787	4.140
15	1.341	1.753	2.131	2.602	2.947	3.733	4.073
16	1.337	1.746	2.120	2.583	2.921	3.686	4.015
17	1.333	1.740	2.110	2.567	2.898	3.646	3.965
18	1.330	1.734	2.101	2.552	2.878	3.610	3.922
19	1.328	1.729	2.093	2.539	2.861	3.579	3.883
20	1.325	1.725	2.086	2.528	2.845	3.552	3.850
21	1.323	1.721	2.080	2.518	2.831	3.527	3.819
22	1.321	1.717	2.074	2.508	2.819	3.505	3.792
23	1.319	1.714	2.069	2.500	2.807	3.485	3.768
24	1.318	1.711	2.064	2.492	2.797	3.467	3.745
25	1.316	1.708	2.060	2.485	2.787	3.450	3.725
26	1.315	1.706	2.056	2.479	2.779	3.435	3.707
27	1.314	1.703	2.052	2.473	2.771	3.421	3.690
28	1.313	1.701	2.048	2.467	2.763	3.408	3.674
29	1.311	1.699	2.045	2.462	2.756	3.396	3.659
30	1.310	1.697	2.042	2.457	2.750	3.385	3.646
35	1.306	1.690	2.030	2.438	2.724	3.340	3.591
40	1.303	1.684	2.021	2.423	2.704	3.307	3.551
45	1.301	1.679	2.014	2.412	2.690	3.281	3.520
50	1.299	1.676	2.009	2.403	2.678	3.261	3.496
100	0.290	1.660	1.984	2.364	2.626	3.174	3.390
∞	1.282	1.645	1.960	2.326	2.576	3.091	3.291

6.13 TRIANGULAR DISTRIBUTION

6.13.1 Properties

$$\text{pdf} \quad f(x) = \begin{cases} 0 & x \leq a \\ 4(x-a)/(b-a)^2 & a < x \leq (a+b)/2 \\ 4(b-x)/(b-a)^2 & (a+b)/2 < x < b \\ 0 & x \geq b \end{cases}$$

$$a < b \in \mathcal{R}$$

mean $\quad \mu = \dfrac{a+b}{2}$

variance $\quad \sigma^2 = \dfrac{(b-a)^2}{24}$

skewness $\quad \beta_1 = 0$

kurtosis $\quad \beta_2 = 12/5$

mgf $\quad m(t) = \dfrac{4(e^{at/2} - e^{bt/2})^2}{(b-a)^2 t^2}$

char function $\quad \phi(t) = -\dfrac{4(e^{ait/2} - e^{bit/2})^2}{(b-a)^2 t^2}$

6.13.2 Probability density function

The probability density function is symmetric about the mean and consists of two line segments.

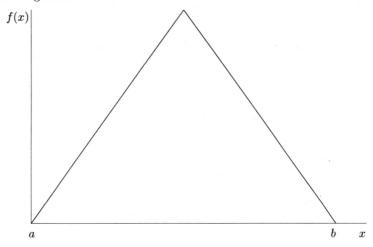

Figure 6.15: Probability density function for a triangular random variable.

6.14 UNIFORM DISTRIBUTION

6.14.1 Properties

pdf	$f(x) = \dfrac{1}{b-a},$	$a \leq x \leq b,\ a < b \in \mathcal{R}$
mean	$\mu = \dfrac{a+b}{2}$	
variance	$\sigma^2 = \dfrac{(b-a)^2}{12}$	
skewness	$\beta_1 = 0$	
kurtosis	$\beta_2 = 9/5$	
mgf	$m(t) = \dfrac{e^{bt} - e^{at}}{(b-a)t}$	
char function	$\phi(t) = \dfrac{e^{bit} - e^{ait}}{(b-a)it}$	

6.14.2 Probability density function

The probability density function is a horizontal line segment between a and b at $1/(b-a)$.

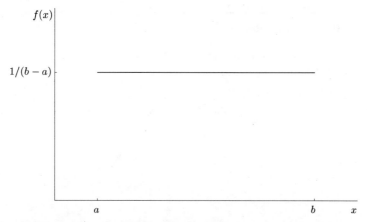

Figure 6.16: Probability density functions for a uniform random variable.

6.14.3 Related distributions

(1) The random variable X has a standard uniform distribution if $a = 0$ and $b = 1$.

(2) If X is a uniform random variable with parameters $a = 0$ and $b = 1$, the random variable $Y = -(\ln X)/\lambda$ has an exponential distribution with parameter λ.

6.15. WEIBULL DISTRIBUTION

(3) Let X_1 and X_2 be independent uniform random variables with parameters $a = 0$ and $b = 1$. The random variable $Y = (X_1 + X_2)/2$ has a triangular distribution with parameters 0 and 1.

(4) If X is a uniform random variable with parameters $a = -\pi/2$ and $b = \pi/2$, then the random variable $Y = \tan X$ has a Cauchy distribution with parameters $a = 0$ and $b = 1$.

6.15 WEIBULL DISTRIBUTION

6.15.1 Properties

$$\text{pdf} \quad f(x) = \frac{\alpha}{\beta^\alpha} x^{\alpha-1} e^{-(x/\beta)^\alpha}$$

$$\text{mean} \quad \mu = \beta \Gamma\left(1 + \frac{1}{\alpha}\right)$$

$$\text{variance} \quad \sigma^2 = \beta^2 \left[\Gamma\left(1 + \frac{2}{\alpha}\right) - \Gamma^2\left(1 + \frac{1}{\alpha}\right)\right]$$

$$\text{skewness} \quad \beta_1 = \frac{2\Gamma^3(1 + \frac{1}{\alpha}) - 3\Gamma(1 + \frac{1}{\alpha})\Gamma(1 + \frac{2}{\alpha}) + \Gamma(1 + \frac{3}{\alpha})}{\left[\Gamma(1 + \frac{2}{\alpha}) - \Gamma^2(1 + \frac{1}{\alpha})\right]^{3/2}}$$

kurtosis $\quad \beta_2 =$

$$\frac{-3\Gamma^4(1 + \frac{1}{\alpha}) + 6\Gamma^2(1 + \frac{1}{\alpha})\Gamma(1 + \frac{2}{\alpha}) - 4\Gamma(1 + \frac{1}{\alpha})\Gamma(1 + \frac{3}{\alpha}) + \Gamma(1 + \frac{4}{\alpha})}{\left[\Gamma(1 + \frac{2}{\alpha}) - \Gamma^2(1 + \frac{1}{\alpha})\right]^2}$$

mgf $\quad m(t) = $ does not exist

char function $\quad \phi(t) = $ does not exist $\cdot$

where $\Gamma(x)$ is the gamma function (see page 204).

6.15.2 Probability density function

The probability density function is skewed to the right. For fixed β the tail becomes lighter and the distribution becomes more bell-shaped as α increases.

6.15.3 Related distributions

Suppose X is a Weibull random variable with parameters α and β.

(1) The random variable X has a standard Weibull distribution if $\beta = 1$.

(2) If $\alpha = 1$ then X has an exponential distribution with parameter $\lambda = 1/\beta$.

(3) The random variable $Y = X^\alpha$ has an exponential distribution with parameter $\lambda = \beta$.

(4) If $\alpha = 2$ then X has a Rayleigh distribution with parameter $\sigma = \beta/\sqrt{2}$.

(5) The random variable $Y = -\alpha \ln(X/\beta)$ has a (standard) extreme-value distribution with parameters $\alpha = 0$ and $\beta = 1$.

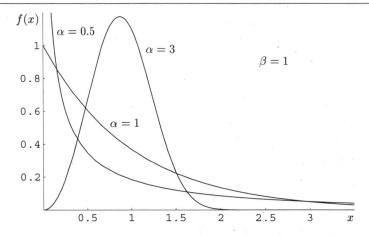

Figure 6.17: Probability density functions for a Weibull random variable.

6.16 RELATIONSHIPS AMONG DISTRIBUTIONS

Figure 6.18 presents some of the relationships among common univariate distributions. The first line of each box is the name of the distribution and the second line lists the parameters that characterize the distribution. The random variable X is used to represent each distribution. The three types of relationships presented in the figure are transformations (independent random variables are assumed) and special cases (both indicated with a solid arrow), and limiting distributions (indicated with a dashed arrow).

6.16.1 Other relationships among distributions

(1) If X_1 has a standard normal distribution, X_2 has a chi-square distribution with ν degrees of freedom, and X_1 and X_2 are independent, then the random variable

$$Y = \frac{X_1}{\sqrt{X_2/\nu}} \qquad (6.11)$$

has a t distribution with ν degrees of freedom.

(2) Let $X_1, X_2, \ldots, X_n$ be independent normal random variables with parameters μ and σ, and define

$$\overline{X} = \frac{1}{n}\sum_{i=1}^{n} X_i \quad \text{and} \quad S^2 = \frac{1}{n}\sum_{i=1}^{n}(X_i - \overline{X})^2. \qquad (6.12)$$

(a) The random variable $Y = nS^2/\sigma^2$ has a chi-square distribution with $n-1$ degrees of freedom.

6.16. RELATIONSHIPS AMONG DISTRIBUTIONS

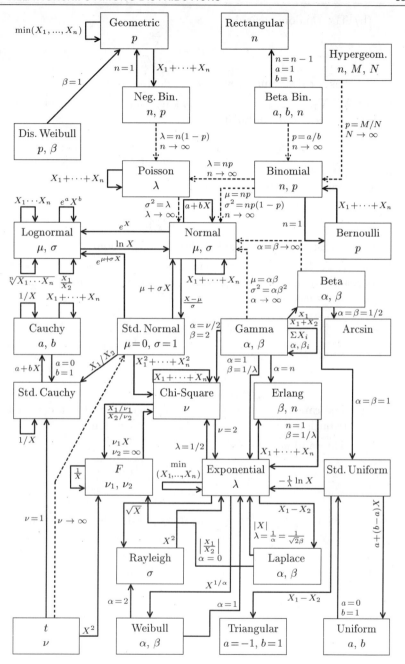

Figure 6.18: Relationships among distributions (see page 110).

(b) The random variable

$$W = \frac{\overline{X} - \mu}{S/\sqrt{n-1}} \qquad (6.13)$$

has a t distribution with $n-1$ degrees of freedom.

(3) Let $X_1, X_2, \ldots, X_n$ be independent normal random variables with parameters μ and σ, and define

$$\overline{X} = \frac{1}{n} \sum_{i=1}^{n} X_i \quad \text{and} \quad S^2 = \frac{1}{n-1} \sum_{i=1}^{n} (X_i - \overline{X})^2. \qquad (6.14)$$

The random variable

$$Y = \frac{\overline{X} - \mu}{S/\sqrt{n}} \qquad (6.15)$$

has a t distribution with $n-1$ degrees of freedom.

(4) Let $X_1, X_2, \ldots, X_{n_1}$ be independent normal random variables with parameters μ_1 and σ, and $Y_1, Y_2, \ldots, Y_{n_2}$ be independent normal random variables with parameters μ_2 and σ. Define

$$\overline{X} = \frac{1}{n_1} \sum_{i=1}^{n_1} X_i \qquad S_1^2 = \frac{1}{n_1} \sum_{i=1}^{n_1} (X_i - \overline{X})^2$$

$$\overline{Y} = \frac{1}{n_2} \sum_{i=1}^{n_2} Y_i \qquad S_2^2 = \frac{1}{n_2} \sum_{i=1}^{n_2} (Y_i - \overline{Y})^2 \qquad (6.16)$$

(a) The random variable $Y = (n_1 S_1^2 + n_2 S_2^2)/\sigma^2$ has a chi-square distribution with $n_1 + n_2 - 2$ degrees of freedom.

(b) The random variable

$$W = \frac{(\overline{X} - \overline{Y}) - (\mu_1 - \mu_2)}{\sqrt{\frac{1}{n_1} + \frac{1}{n_2}} \sqrt{\frac{n_1 S_1^2 + n_2 S_2^2}{n_1 + n_2 - 2}}} \qquad (6.17)$$

has a t distribution with $n_1 + n_2 - 2$ degrees of freedom.

(5) Let $X_1, X_2, \ldots, X_{n_1}$ be independent normal random variables with parameters μ_1 and σ_1, and $Y_1, Y_2, \ldots, Y_{n_2}$ be independent normal random variables with parameters μ_2 and σ_2. Define

$$\overline{X} = \frac{1}{n_1} \sum_{i=1}^{n_1} X_i \qquad S_1^2 = \frac{1}{n_1} \sum_{i=1}^{n_1} (X_i - \overline{X})^2$$

$$\overline{Y} = \frac{1}{n_2} \sum_{i=1}^{n_2} Y_i \qquad S_2^2 = \frac{1}{n_2} \sum_{i=1}^{n_2} (Y_i - \overline{Y})^2 \qquad (6.18)$$

The random variable

$$Y = \frac{n_1 S_1^2}{(n_1 - 1)\sigma_1^2} \bigg/ \frac{n_2 S_2^2}{(n_2 - 1)\sigma_2^2} \qquad (6.19)$$

6.16. RELATIONSHIPS AMONG DISTRIBUTIONS

has an F distribution with n_1 and n_2 degrees of freedom.

(6) Let X_1 be a normal random variable with parameters $\mu = \lambda$ and $\sigma = 1$, X_2 a chi-square random variable with parameter ν, and X_1 and X_2 be independent. The random variable $Y = X_1/\sqrt{X_2/\nu}$ has a noncentral t distribution with parameters ν and λ.

(7) Let X be a continuous random variable with cumulative distribution function $F(x)$.

  (a) The random variable $Y = F(X)$ has a (standard) uniform distribution with parameters $a = 0$ and $b = 1$.

  (b) The random variable $Y = -\ln[1 - F(X)]$ has a (standard) exponential distribution with parameter $\lambda = 1$.

Let X be a continuous random variable with probability density function $f(x)$. The random variable $Y = |X|$ has probability density function $g(y)$ given by

$$g(y) = \begin{cases} f(y) + f(-y) & \text{if } y > 0 \\ 0 & \text{elsewhere} \end{cases} \qquad (6.20)$$

If X has a standard normal distribution ($\mu = 0$, $\sigma = 1$) then $g(y) = 2f(y)$.

CHAPTER 7

Standard Normal Distribution

7.1 THE PROBABILITY DENSITY FUNCTION AND RELATED FUNCTIONS

Let Z be a standard normal random variable ($\mu = 0$, $\sigma = 1$). The probability density function is given by

$$f(z) = \frac{1}{\sqrt{2\pi}} e^{-z^2/2}. \tag{7.1}$$

The following tables contain values for:

(1) $f(z)$

(2) $F(z) = \text{Prob}\,[Z \leq z] = \int_{-\infty}^{z} \frac{1}{\sqrt{2\pi}} e^{-t^2/2}\, dt.$

$\quad\ $ = the cumulative distribution function

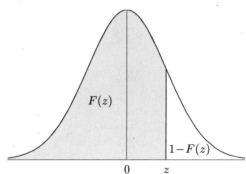

Figure 7.1: Cumulative distribution function for a standard normal random variable.

Note:

(1) For all z, $f(-z) = f(z)$.

(2) For all z, $F(-z) = 1 - F(z)$.

(3) For all z, $\text{Prob}\,[|Z| \leq z] = F(z) - F(-z)$

(4) For all z, $\text{Prob}\,[|Z| \geq z] = 1 - F(z) + F(-z)$

(5) The function $\Phi(z) = F(z)$ is often used to represent the normal distribution function.

(6) $f'(x) = -\frac{x}{\sqrt{2\pi}}e^{-x^2/2} = -x\,f(x)$

(7) $f''(x) = \left(x^2 - 1\right) f(x)$

(8) $f'''(x) = \left(3x - x^3\right) f(x)$

(9) $f^{(4)}(x) = \left(x^4 - 6x^2 + 3\right) f(x)$

(10) For large values of x:
$$\left[\frac{e^{-x^2/2}}{\sqrt{2\pi}} \left(\frac{1}{x} - \frac{1}{x^3}\right)\right] < 1 - \Phi(x) < \left[\frac{e^{-x^2/2}}{\sqrt{2\pi}} \left(\frac{1}{x}\right)\right] \quad (7.2)$$

(11) Order statistics for the normal distribution may be found in section 4.6.8 on page 51.

7.1. DENSITY FUNCTION AND RELATED FUNCTIONS

Normal distribution

z	$f(z)$	$F(z)$	$1-F(z)$	z	$f(z)$	$F(z)$	$1-F(z)$
-4.00	0.0001	0.0000	1.0000	-3.50	0.0009	0.0002	0.9998
-3.99	0.0001	0.0000	1.0000	-3.49	0.0009	0.0002	0.9998
-3.98	0.0001	0.0000	1.0000	-3.48	0.0009	0.0003	0.9998
-3.97	0.0001	0.0000	1.0000	-3.47	0.0010	0.0003	0.9997
-3.96	0.0002	0.0000	1.0000	-3.46	0.0010	0.0003	0.9997
-3.95	0.0002	0.0000	1.0000	-3.45	0.0010	0.0003	0.9997
-3.94	0.0002	0.0000	1.0000	-3.44	0.0011	0.0003	0.9997
-3.93	0.0002	0.0000	1.0000	-3.43	0.0011	0.0003	0.9997
-3.92	0.0002	0.0000	1.0000	-3.42	0.0011	0.0003	0.9997
-3.91	0.0002	0.0001	1.0000	-3.41	0.0012	0.0003	0.9997
-3.90	0.0002	0.0001	1.0000	-3.40	0.0012	0.0003	0.9997
-3.89	0.0002	0.0001	1.0000	-3.39	0.0013	0.0003	0.9997
-3.88	0.0002	0.0001	1.0000	-3.38	0.0013	0.0004	0.9996
-3.87	0.0002	0.0001	1.0000	-3.37	0.0014	0.0004	0.9996
-3.86	0.0002	0.0001	0.9999	-3.36	0.0014	0.0004	0.9996
-3.85	0.0002	0.0001	0.9999	-3.35	0.0015	0.0004	0.9996
-3.84	0.0003	0.0001	0.9999	-3.34	0.0015	0.0004	0.9996
-3.83	0.0003	0.0001	0.9999	-3.33	0.0016	0.0004	0.9996
-3.82	0.0003	0.0001	0.9999	-3.32	0.0016	0.0004	0.9996
-3.81	0.0003	0.0001	0.9999	-3.31	0.0017	0.0005	0.9995
-3.80	0.0003	0.0001	0.9999	-3.30	0.0017	0.0005	0.9995
-3.79	0.0003	0.0001	0.9999	-3.29	0.0018	0.0005	0.9995
-3.78	0.0003	0.0001	0.9999	-3.28	0.0018	0.0005	0.9995
-3.77	0.0003	0.0001	0.9999	-3.27	0.0019	0.0005	0.9995
-3.76	0.0003	0.0001	0.9999	-3.26	0.0020	0.0006	0.9994
-3.75	0.0003	0.0001	0.9999	-3.25	0.0020	0.0006	0.9994
-3.74	0.0004	0.0001	0.9999	-3.24	0.0021	0.0006	0.9994
-3.73	0.0004	0.0001	0.9999	-3.23	0.0022	0.0006	0.9994
-3.72	0.0004	0.0001	0.9999	-3.22	0.0022	0.0006	0.9994
-3.71	0.0004	0.0001	0.9999	-3.21	0.0023	0.0007	0.9993
-3.70	0.0004	0.0001	0.9999	-3.20	0.0024	0.0007	0.9993
-3.69	0.0004	0.0001	0.9999	-3.19	0.0025	0.0007	0.9993
-3.68	0.0005	0.0001	0.9999	-3.18	0.0025	0.0007	0.9993
-3.67	0.0005	0.0001	0.9999	-3.17	0.0026	0.0008	0.9992
-3.66	0.0005	0.0001	0.9999	-3.16	0.0027	0.0008	0.9992
-3.65	0.0005	0.0001	0.9999	-3.15	0.0028	0.0008	0.9992
-3.64	0.0005	0.0001	0.9999	-3.14	0.0029	0.0008	0.9992
-3.63	0.0006	0.0001	0.9999	-3.13	0.0030	0.0009	0.9991
-3.62	0.0006	0.0001	0.9999	-3.12	0.0031	0.0009	0.9991
-3.61	0.0006	0.0001	0.9999	-3.11	0.0032	0.0009	0.9991
-3.60	0.0006	0.0002	0.9998	-3.10	0.0033	0.0010	0.9990
-3.59	0.0006	0.0002	0.9998	-3.09	0.0034	0.0010	0.9990
-3.58	0.0007	0.0002	0.9998	-3.08	0.0035	0.0010	0.9990
-3.57	0.0007	0.0002	0.9998	-3.07	0.0036	0.0011	0.9989
-3.56	0.0007	0.0002	0.9998	-3.06	0.0037	0.0011	0.9989
-3.55	0.0007	0.0002	0.9998	-3.05	0.0038	0.0011	0.9989
-3.54	0.0008	0.0002	0.9998	-3.04	0.0039	0.0012	0.9988
-3.53	0.0008	0.0002	0.9998	-3.03	0.0040	0.0012	0.9988
-3.52	0.0008	0.0002	0.9998	-3.02	0.0042	0.0013	0.9987
-3.51	0.0008	0.0002	0.9998	-3.01	0.0043	0.0013	0.9987
-3.50	0.0009	0.0002	0.9998	-3.00	0.0044	0.0014	0.9987

Normal distribution

z	$f(z)$	$F(z)$	$1-F(z)$	z	$f(z)$	$F(z)$	$1-F(z)$
−3.00	0.0044	0.0014	0.9987	−2.50	0.0175	0.0062	0.9938
−2.99	0.0046	0.0014	0.9986	−2.49	0.0180	0.0064	0.9936
−2.98	0.0047	0.0014	0.9986	−2.48	0.0184	0.0066	0.9934
−2.97	0.0049	0.0015	0.9985	−2.47	0.0189	0.0068	0.9932
−2.96	0.0050	0.0015	0.9985	−2.46	0.0194	0.0069	0.9930
−2.95	0.0051	0.0016	0.9984	−2.45	0.0198	0.0071	0.9929
−2.94	0.0053	0.0016	0.9984	−2.44	0.0203	0.0073	0.9927
−2.93	0.0054	0.0017	0.9983	−2.43	0.0208	0.0076	0.9925
−2.92	0.0056	0.0018	0.9982	−2.42	0.0213	0.0078	0.9922
−2.91	0.0058	0.0018	0.9982	−2.41	0.0219	0.0080	0.9920
−2.90	0.0060	0.0019	0.9981	−2.40	0.0224	0.0082	0.9918
−2.89	0.0061	0.0019	0.9981	−2.39	0.0229	0.0084	0.9916
−2.88	0.0063	0.0020	0.9980	−2.38	0.0235	0.0087	0.9913
−2.87	0.0065	0.0021	0.9980	−2.37	0.0241	0.0089	0.9911
−2.86	0.0067	0.0021	0.9979	−2.36	0.0246	0.0091	0.9909
−2.85	0.0069	0.0022	0.9978	−2.35	0.0252	0.0094	0.9906
−2.84	0.0071	0.0023	0.9977	−2.34	0.0258	0.0096	0.9904
−2.83	0.0073	0.0023	0.9977	−2.33	0.0264	0.0099	0.9901
−2.82	0.0075	0.0024	0.9976	−2.32	0.0271	0.0102	0.9898
−2.81	0.0077	0.0025	0.9975	−2.31	0.0277	0.0104	0.9896
−2.80	0.0079	0.0026	0.9974	−2.30	0.0283	0.0107	0.9893
−2.79	0.0081	0.0026	0.9974	−2.29	0.0290	0.0110	0.9890
−2.78	0.0084	0.0027	0.9973	−2.28	0.0296	0.0113	0.9887
−2.77	0.0086	0.0028	0.9972	−2.27	0.0303	0.0116	0.9884
−2.76	0.0089	0.0029	0.9971	−2.26	0.0310	0.0119	0.9881
−2.75	0.0091	0.0030	0.9970	−2.25	0.0317	0.0122	0.9878
−2.74	0.0094	0.0031	0.9969	−2.24	0.0325	0.0126	0.9875
−2.73	0.0096	0.0032	0.9968	−2.23	0.0332	0.0129	0.9871
−2.72	0.0099	0.0033	0.9967	−2.22	0.0339	0.0132	0.9868
−2.71	0.0101	0.0034	0.9966	−2.21	0.0347	0.0135	0.9865
−2.70	0.0104	0.0035	0.9965	−2.20	0.0355	0.0139	0.9861
−2.69	0.0107	0.0036	0.9964	−2.19	0.0363	0.0143	0.9857
−2.68	0.0110	0.0037	0.9963	−2.18	0.0371	0.0146	0.9854
−2.67	0.0113	0.0038	0.9962	−2.17	0.0379	0.0150	0.9850
−2.66	0.0116	0.0039	0.9961	−2.16	0.0387	0.0154	0.9846
−2.65	0.0119	0.0040	0.9960	−2.15	0.0396	0.0158	0.9842
−2.64	0.0122	0.0042	0.9959	−2.14	0.0404	0.0162	0.9838
−2.63	0.0126	0.0043	0.9957	−2.13	0.0413	0.0166	0.9834
−2.62	0.0129	0.0044	0.9956	−2.12	0.0422	0.0170	0.9830
−2.61	0.0132	0.0045	0.9955	−2.11	0.0431	0.0174	0.9826
−2.60	0.0136	0.0047	0.9953	−2.10	0.0440	0.0179	0.9821
−2.59	0.0139	0.0048	0.9952	−2.09	0.0449	0.0183	0.9817
−2.58	0.0143	0.0049	0.9951	−2.08	0.0459	0.0188	0.9812
−2.57	0.0147	0.0051	0.9949	−2.07	0.0468	0.0192	0.9808
−2.56	0.0151	0.0052	0.9948	−2.06	0.0478	0.0197	0.9803
−2.55	0.0155	0.0054	0.9946	−2.05	0.0488	0.0202	0.9798
−2.54	0.0158	0.0055	0.9945	−2.04	0.0498	0.0207	0.9793
−2.53	0.0163	0.0057	0.9943	−2.03	0.0508	0.0212	0.9788
−2.52	0.0167	0.0059	0.9941	−2.02	0.0519	0.0217	0.9783
−2.51	0.0171	0.0060	0.9940	−2.01	0.0529	0.0222	0.9778
−2.50	0.0175	0.0062	0.9938	−2.00	0.0540	0.0227	0.9772

7.1. DENSITY FUNCTION AND RELATED FUNCTIONS

Normal distribution

z	$f(z)$	$F(z)$	$1-F(z)$	z	$f(z)$	$F(z)$	$1-F(z)$
-2.00	0.0540	0.0227	0.9772	-1.50	0.1295	0.0668	0.9332
-1.99	0.0551	0.0233	0.9767	-1.49	0.1315	0.0681	0.9319
-1.98	0.0562	0.0238	0.9761	-1.48	0.1334	0.0694	0.9306
-1.97	0.0573	0.0244	0.9756	-1.47	0.1354	0.0708	0.9292
-1.96	0.0584	0.0250	0.9750	-1.46	0.1374	0.0722	0.9278
-1.95	0.0596	0.0256	0.9744	-1.45	0.1394	0.0735	0.9265
-1.94	0.0608	0.0262	0.9738	-1.44	0.1415	0.0749	0.9251
-1.93	0.0619	0.0268	0.9732	-1.43	0.1435	0.0764	0.9236
-1.92	0.0632	0.0274	0.9726	-1.42	0.1456	0.0778	0.9222
-1.91	0.0644	0.0281	0.9719	-1.41	0.1476	0.0793	0.9207
-1.90	0.0656	0.0287	0.9713	-1.40	0.1497	0.0808	0.9192
-1.89	0.0669	0.0294	0.9706	-1.39	0.1518	0.0823	0.9177
-1.88	0.0681	0.0301	0.9699	-1.38	0.1540	0.0838	0.9162
-1.87	0.0694	0.0307	0.9693	-1.37	0.1561	0.0853	0.9147
-1.86	0.0707	0.0314	0.9686	-1.36	0.1582	0.0869	0.9131
-1.85	0.0721	0.0322	0.9678	-1.35	0.1604	0.0885	0.9115
-1.84	0.0734	0.0329	0.9671	-1.34	0.1626	0.0901	0.9099
-1.83	0.0748	0.0336	0.9664	-1.33	0.1647	0.0918	0.9082
-1.82	0.0761	0.0344	0.9656	-1.32	0.1669	0.0934	0.9066
-1.81	0.0775	0.0352	0.9648	-1.31	0.1691	0.0951	0.9049
-1.80	0.0790	0.0359	0.9641	-1.30	0.1714	0.0968	0.9032
-1.79	0.0804	0.0367	0.9633	-1.29	0.1736	0.0985	0.9015
-1.78	0.0818	0.0375	0.9625	-1.28	0.1759	0.1003	0.8997
-1.77	0.0833	0.0384	0.9616	-1.27	0.1781	0.1020	0.8980
-1.76	0.0848	0.0392	0.9608	-1.26	0.1804	0.1038	0.8962
-1.75	0.0863	0.0401	0.9599	-1.25	0.1827	0.1056	0.8943
-1.74	0.0878	0.0409	0.9591	-1.24	0.1849	0.1075	0.8925
-1.73	0.0893	0.0418	0.9582	-1.23	0.1872	0.1094	0.8907
-1.72	0.0909	0.0427	0.9573	-1.22	0.1895	0.1112	0.8888
-1.71	0.0925	0.0436	0.9564	-1.21	0.1919	0.1131	0.8869
-1.70	0.0940	0.0446	0.9554	-1.20	0.1942	0.1151	0.8849
-1.69	0.0957	0.0455	0.9545	-1.19	0.1965	0.1170	0.8830
-1.68	0.0973	0.0465	0.9535	-1.18	0.1989	0.1190	0.8810
-1.67	0.0989	0.0475	0.9525	-1.17	0.2012	0.1210	0.8790
-1.66	0.1006	0.0485	0.9515	-1.16	0.2036	0.1230	0.8770
-1.65	0.1023	0.0495	0.9505	-1.15	0.2059	0.1251	0.8749
-1.64	0.1040	0.0505	0.9495	-1.14	0.2083	0.1271	0.8729
-1.63	0.1057	0.0515	0.9485	-1.13	0.2107	0.1292	0.8708
-1.62	0.1074	0.0526	0.9474	-1.12	0.2131	0.1314	0.8686
-1.61	0.1091	0.0537	0.9463	-1.11	0.2155	0.1335	0.8665
-1.60	0.1109	0.0548	0.9452	-1.10	0.2178	0.1357	0.8643
-1.59	0.1127	0.0559	0.9441	-1.09	0.2203	0.1379	0.8621
-1.58	0.1145	0.0570	0.9429	-1.08	0.2226	0.1401	0.8599
-1.57	0.1163	0.0582	0.9418	-1.07	0.2251	0.1423	0.8577
-1.56	0.1182	0.0594	0.9406	-1.06	0.2275	0.1446	0.8554
-1.55	0.1200	0.0606	0.9394	-1.05	0.2299	0.1469	0.8531
-1.54	0.1219	0.0618	0.9382	-1.04	0.2323	0.1492	0.8508
-1.53	0.1238	0.0630	0.9370	-1.03	0.2347	0.1515	0.8485
-1.52	0.1257	0.0643	0.9357	-1.02	0.2371	0.1539	0.8461
-1.51	0.1276	0.0655	0.9345	-1.01	0.2396	0.1563	0.8438
-1.50	0.1295	0.0668	0.9332	-1.00	0.2420	0.1587	0.8413

Normal distribution

z	$f(z)$	$F(z)$	$1 - F(z)$	z	$f(z)$	$F(z)$	$1 - F(z)$
−1.00	0.2420	0.1587	0.8413	−0.50	0.3521	0.3085	0.6915
−0.99	0.2444	0.1611	0.8389	−0.49	0.3538	0.3121	0.6879
−0.98	0.2468	0.1635	0.8365	−0.48	0.3555	0.3156	0.6844
−0.97	0.2492	0.1660	0.8340	−0.47	0.3572	0.3192	0.6808
−0.96	0.2516	0.1685	0.8315	−0.46	0.3589	0.3228	0.6772
−0.95	0.2541	0.1711	0.8289	−0.45	0.3605	0.3264	0.6736
−0.94	0.2565	0.1736	0.8264	−0.44	0.3621	0.3300	0.6700
−0.93	0.2589	0.1762	0.8238	−0.43	0.3637	0.3336	0.6664
−0.92	0.2613	0.1788	0.8212	−0.42	0.3653	0.3372	0.6628
−0.91	0.2637	0.1814	0.8186	−0.41	0.3668	0.3409	0.6591
−0.90	0.2661	0.1841	0.8159	−0.40	0.3683	0.3446	0.6554
−0.89	0.2685	0.1867	0.8133	−0.39	0.3697	0.3483	0.6517
−0.88	0.2709	0.1894	0.8106	−0.38	0.3711	0.3520	0.6480
−0.87	0.2732	0.1921	0.8078	−0.37	0.3725	0.3557	0.6443
−0.86	0.2756	0.1949	0.8051	−0.36	0.3739	0.3594	0.6406
−0.85	0.2780	0.1977	0.8023	−0.35	0.3752	0.3632	0.6368
−0.84	0.2803	0.2004	0.7995	−0.34	0.3765	0.3669	0.6331
−0.83	0.2827	0.2033	0.7967	−0.33	0.3778	0.3707	0.6293
−0.82	0.2850	0.2061	0.7939	−0.32	0.3790	0.3745	0.6255
−0.81	0.2874	0.2090	0.7910	−0.31	0.3802	0.3783	0.6217
−0.80	0.2897	0.2119	0.7881	−0.30	0.3814	0.3821	0.6179
−0.79	0.2920	0.2148	0.7852	−0.29	0.3825	0.3859	0.6141
−0.78	0.2943	0.2177	0.7823	−0.28	0.3836	0.3897	0.6103
−0.77	0.2966	0.2207	0.7793	−0.27	0.3847	0.3936	0.6064
−0.76	0.2989	0.2236	0.7764	−0.26	0.3857	0.3974	0.6026
−0.75	0.3011	0.2266	0.7734	−0.25	0.3867	0.4013	0.5987
−0.74	0.3034	0.2296	0.7703	−0.24	0.3876	0.4052	0.5948
−0.73	0.3056	0.2327	0.7673	−0.23	0.3885	0.4091	0.5909
−0.72	0.3079	0.2358	0.7642	−0.22	0.3894	0.4129	0.5871
−0.71	0.3101	0.2389	0.7611	−0.21	0.3902	0.4168	0.5832
−0.70	0.3123	0.2420	0.7580	−0.20	0.3910	0.4207	0.5793
−0.69	0.3144	0.2451	0.7549	−0.19	0.3918	0.4247	0.5754
−0.68	0.3166	0.2482	0.7518	−0.18	0.3925	0.4286	0.5714
−0.67	0.3187	0.2514	0.7486	−0.17	0.3932	0.4325	0.5675
−0.66	0.3209	0.2546	0.7454	−0.16	0.3939	0.4364	0.5636
−0.65	0.3230	0.2579	0.7421	−0.15	0.3945	0.4404	0.5596
−0.64	0.3251	0.2611	0.7389	−0.14	0.3951	0.4443	0.5557
−0.63	0.3271	0.2643	0.7357	−0.13	0.3956	0.4483	0.5517
−0.62	0.3292	0.2676	0.7324	−0.12	0.3961	0.4522	0.5478
−0.61	0.3312	0.2709	0.7291	−0.11	0.3965	0.4562	0.5438
−0.60	0.3332	0.2742	0.7258	−0.10	0.3970	0.4602	0.5398
−0.59	0.3352	0.2776	0.7224	−0.09	0.3973	0.4641	0.5359
−0.58	0.3372	0.2810	0.7190	−0.08	0.3977	0.4681	0.5319
−0.57	0.3391	0.2843	0.7157	−0.07	0.3980	0.4721	0.5279
−0.56	0.3411	0.2877	0.7123	−0.06	0.3982	0.4761	0.5239
−0.55	0.3429	0.2912	0.7088	−0.05	0.3984	0.4801	0.5199
−0.54	0.3448	0.2946	0.7054	−0.04	0.3986	0.4840	0.5160
−0.53	0.3467	0.2981	0.7019	−0.03	0.3988	0.4880	0.5120
−0.52	0.3485	0.3015	0.6985	−0.02	0.3989	0.4920	0.5080
−0.51	0.3503	0.3050	0.6950	−0.01	0.3989	0.4960	0.5040
−0.50	0.3521	0.3085	0.6915	0.00	0.3989	0.5000	0.5000

7.1. DENSITY FUNCTION AND RELATED FUNCTIONS

Normal distribution

z	$f(z)$	$F(z)$	$1 - F(z)$	z	$f(z)$	$F(z)$	$1 - F(z)$
0.00	0.3989	0.5000	0.5000	0.50	0.3521	0.6915	0.3085
0.01	0.3989	0.5040	0.4960	0.51	0.3503	0.6950	0.3050
0.02	0.3989	0.5080	0.4920	0.52	0.3485	0.6985	0.3015
0.03	0.3988	0.5120	0.4880	0.53	0.3467	0.7019	0.2981
0.04	0.3986	0.5160	0.4840	0.54	0.3448	0.7054	0.2946
0.05	0.3984	0.5199	0.4801	0.55	0.3429	0.7088	0.2912
0.06	0.3982	0.5239	0.4761	0.56	0.3411	0.7123	0.2877
0.07	0.3980	0.5279	0.4721	0.57	0.3391	0.7157	0.2843
0.08	0.3977	0.5319	0.4681	0.58	0.3372	0.7190	0.2810
0.09	0.3973	0.5359	0.4641	0.59	0.3352	0.7224	0.2776
0.10	0.3970	0.5398	0.4602	0.60	0.3332	0.7258	0.2742
0.11	0.3965	0.5438	0.4562	0.61	0.3312	0.7291	0.2709
0.12	0.3961	0.5478	0.4522	0.62	0.3292	0.7324	0.2676
0.13	0.3956	0.5517	0.4483	0.63	0.3271	0.7357	0.2643
0.14	0.3951	0.5557	0.4443	0.64	0.3251	0.7389	0.2611
0.15	0.3945	0.5596	0.4404	0.65	0.3230	0.7421	0.2579
0.16	0.3939	0.5636	0.4364	0.66	0.3209	0.7454	0.2546
0.17	0.3932	0.5675	0.4325	0.67	0.3187	0.7486	0.2514
0.18	0.3925	0.5714	0.4286	0.68	0.3166	0.7518	0.2482
0.19	0.3918	0.5754	0.4247	0.69	0.3144	0.7549	0.2451
0.20	0.3910	0.5793	0.4207	0.70	0.3123	0.7580	0.2420
0.21	0.3902	0.5832	0.4168	0.71	0.3101	0.7611	0.2389
0.22	0.3894	0.5871	0.4129	0.72	0.3079	0.7642	0.2358
0.23	0.3885	0.5909	0.4091	0.73	0.3056	0.7673	0.2327
0.24	0.3876	0.5948	0.4052	0.74	0.3034	0.7703	0.2296
0.25	0.3867	0.5987	0.4013	0.75	0.3011	0.7734	0.2266
0.26	0.3857	0.6026	0.3974	0.76	0.2989	0.7764	0.2236
0.27	0.3847	0.6064	0.3936	0.77	0.2966	0.7793	0.2207
0.28	0.3836	0.6103	0.3897	0.78	0.2943	0.7823	0.2177
0.29	0.3825	0.6141	0.3859	0.79	0.2920	0.7852	0.2148
0.30	0.3814	0.6179	0.3821	0.80	0.2897	0.7881	0.2119
0.31	0.3802	0.6217	0.3783	0.81	0.2874	0.7910	0.2090
0.32	0.3790	0.6255	0.3745	0.82	0.2850	0.7939	0.2061
0.33	0.3778	0.6293	0.3707	0.83	0.2827	0.7967	0.2033
0.34	0.3765	0.6331	0.3669	0.84	0.2803	0.7995	0.2004
0.35	0.3752	0.6368	0.3632	0.85	0.2780	0.8023	0.1977
0.36	0.3739	0.6406	0.3594	0.86	0.2756	0.8051	0.1949
0.37	0.3725	0.6443	0.3557	0.87	0.2732	0.8078	0.1921
0.38	0.3711	0.6480	0.3520	0.88	0.2709	0.8106	0.1894
0.39	0.3697	0.6517	0.3483	0.89	0.2685	0.8133	0.1867
0.40	0.3683	0.6554	0.3446	0.90	0.2661	0.8159	0.1841
0.41	0.3668	0.6591	0.3409	0.91	0.2637	0.8186	0.1814
0.42	0.3653	0.6628	0.3372	0.92	0.2613	0.8212	0.1788
0.43	0.3637	0.6664	0.3336	0.93	0.2589	0.8238	0.1762
0.44	0.3621	0.6700	0.3300	0.94	0.2565	0.8264	0.1736
0.45	0.3605	0.6736	0.3264	0.95	0.2541	0.8289	0.1711
0.46	0.3589	0.6772	0.3228	0.96	0.2516	0.8315	0.1685
0.47	0.3572	0.6808	0.3192	0.97	0.2492	0.8340	0.1660
0.48	0.3555	0.6844	0.3156	0.98	0.2468	0.8365	0.1635
0.49	0.3538	0.6879	0.3121	0.99	0.2444	0.8389	0.1611
0.50	0.3521	0.6915	0.3085	1.00	0.2420	0.8413	0.1587

Normal distribution

z	$f(z)$	$F(z)$	$1-F(z)$	z	$f(z)$	$F(z)$	$1-F(z)$
1.00	0.2420	0.8413	0.1587	1.50	0.1295	0.9332	0.0668
1.01	0.2396	0.8438	0.1563	1.51	0.1276	0.9345	0.0655
1.02	0.2371	0.8461	0.1539	1.52	0.1257	0.9357	0.0643
1.03	0.2347	0.8485	0.1515	1.53	0.1238	0.9370	0.0630
1.04	0.2323	0.8508	0.1492	1.54	0.1219	0.9382	0.0618
1.05	0.2299	0.8531	0.1469	1.55	0.1200	0.9394	0.0606
1.06	0.2275	0.8554	0.1446	1.56	0.1182	0.9406	0.0594
1.07	0.2251	0.8577	0.1423	1.57	0.1163	0.9418	0.0582
1.08	0.2226	0.8599	0.1401	1.58	0.1145	0.9429	0.0570
1.09	0.2203	0.8621	0.1379	1.59	0.1127	0.9441	0.0559
1.10	0.2178	0.8643	0.1357	1.60	0.1109	0.9452	0.0548
1.11	0.2155	0.8665	0.1335	1.61	0.1091	0.9463	0.0537
1.12	0.2131	0.8686	0.1314	1.62	0.1074	0.9474	0.0526
1.13	0.2107	0.8708	0.1292	1.63	0.1057	0.9485	0.0515
1.14	0.2083	0.8729	0.1271	1.64	0.1040	0.9495	0.0505
1.15	0.2059	0.8749	0.1251	1.65	0.1023	0.9505	0.0495
1.16	0.2036	0.8770	0.1230	1.66	0.1006	0.9515	0.0485
1.17	0.2012	0.8790	0.1210	1.67	0.0989	0.9525	0.0475
1.18	0.1989	0.8810	0.1190	1.68	0.0973	0.9535	0.0465
1.19	0.1965	0.8830	0.1170	1.69	0.0957	0.9545	0.0455
1.20	0.1942	0.8849	0.1151	1.70	0.0940	0.9554	0.0446
1.21	0.1919	0.8869	0.1131	1.71	0.0925	0.9564	0.0436
1.22	0.1895	0.8888	0.1112	1.72	0.0909	0.9573	0.0427
1.23	0.1872	0.8907	0.1094	1.73	0.0893	0.9582	0.0418
1.24	0.1849	0.8925	0.1075	1.74	0.0878	0.9591	0.0409
1.25	0.1827	0.8943	0.1056	1.75	0.0863	0.9599	0.0401
1.26	0.1804	0.8962	0.1038	1.76	0.0848	0.9608	0.0392
1.27	0.1781	0.8980	0.1020	1.77	0.0833	0.9616	0.0384
1.28	0.1759	0.8997	0.1003	1.78	0.0818	0.9625	0.0375
1.29	0.1736	0.9015	0.0985	1.79	0.0804	0.9633	0.0367
1.30	0.1714	0.9032	0.0968	1.80	0.0790	0.9641	0.0359
1.31	0.1691	0.9049	0.0951	1.81	0.0775	0.9648	0.0352
1.32	0.1669	0.9066	0.0934	1.82	0.0761	0.9656	0.0344
1.33	0.1647	0.9082	0.0918	1.83	0.0748	0.9664	0.0336
1.34	0.1626	0.9099	0.0901	1.84	0.0734	0.9671	0.0329
1.35	0.1604	0.9115	0.0885	1.85	0.0721	0.9678	0.0322
1.36	0.1582	0.9131	0.0869	1.86	0.0707	0.9686	0.0314
1.37	0.1561	0.9147	0.0853	1.87	0.0694	0.9693	0.0307
1.38	0.1540	0.9162	0.0838	1.88	0.0681	0.9699	0.0301
1.39	0.1518	0.9177	0.0823	1.89	0.0669	0.9706	0.0294
1.40	0.1497	0.9192	0.0808	1.90	0.0656	0.9713	0.0287
1.41	0.1476	0.9207	0.0793	1.91	0.0644	0.9719	0.0281
1.42	0.1456	0.9222	0.0778	1.92	0.0632	0.9726	0.0274
1.43	0.1435	0.9236	0.0764	1.93	0.0619	0.9732	0.0268
1.44	0.1415	0.9251	0.0749	1.94	0.0608	0.9738	0.0262
1.45	0.1394	0.9265	0.0735	1.95	0.0596	0.9744	0.0256
1.46	0.1374	0.9278	0.0722	1.96	0.0584	0.9750	0.0250
1.47	0.1354	0.9292	0.0708	1.97	0.0573	0.9756	0.0244
1.48	0.1334	0.9306	0.0694	1.98	0.0562	0.9761	0.0238
1.49	0.1315	0.9319	0.0681	1.99	0.0551	0.9767	0.0233
1.50	0.1295	0.9332	0.0668	2.00	0.0540	0.9772	0.0227

7.1. DENSITY FUNCTION AND RELATED FUNCTIONS

Normal distribution

z	$f(z)$	$F(z)$	$1-F(z)$	z	$f(z)$	$F(z)$	$1-F(z)$
2.00	0.0540	0.9772	0.0227	2.50	0.0175	0.9938	0.0062
2.01	0.0529	0.9778	0.0222	2.51	0.0171	0.9940	0.0060
2.02	0.0519	0.9783	0.0217	2.52	0.0167	0.9941	0.0059
2.03	0.0508	0.9788	0.0212	2.53	0.0163	0.9943	0.0057
2.04	0.0498	0.9793	0.0207	2.54	0.0158	0.9945	0.0055
2.05	0.0488	0.9798	0.0202	2.55	0.0155	0.9946	0.0054
2.06	0.0478	0.9803	0.0197	2.56	0.0151	0.9948	0.0052
2.07	0.0468	0.9808	0.0192	2.57	0.0147	0.9949	0.0051
2.08	0.0459	0.9812	0.0188	2.58	0.0143	0.9951	0.0049
2.09	0.0449	0.9817	0.0183	2.59	0.0139	0.9952	0.0048
2.10	0.0440	0.9821	0.0179	2.60	0.0136	0.9953	0.0047
2.11	0.0431	0.9826	0.0174	2.61	0.0132	0.9955	0.0045
2.12	0.0422	0.9830	0.0170	2.62	0.0129	0.9956	0.0044
2.13	0.0413	0.9834	0.0166	2.63	0.0126	0.9957	0.0043
2.14	0.0404	0.9838	0.0162	2.64	0.0122	0.9959	0.0042
2.15	0.0396	0.9842	0.0158	2.65	0.0119	0.9960	0.0040
2.16	0.0387	0.9846	0.0154	2.66	0.0116	0.9961	0.0039
2.17	0.0379	0.9850	0.0150	2.67	0.0113	0.9962	0.0038
2.18	0.0371	0.9854	0.0146	2.68	0.0110	0.9963	0.0037
2.19	0.0363	0.9857	0.0143	2.69	0.0107	0.9964	0.0036
2.20	0.0355	0.9861	0.0139	2.70	0.0104	0.9965	0.0035
2.21	0.0347	0.9865	0.0135	2.71	0.0101	0.9966	0.0034
2.22	0.0339	0.9868	0.0132	2.72	0.0099	0.9967	0.0033
2.23	0.0332	0.9871	0.0129	2.73	0.0096	0.9968	0.0032
2.24	0.0325	0.9875	0.0126	2.74	0.0094	0.9969	0.0031
2.25	0.0317	0.9878	0.0122	2.75	0.0091	0.9970	0.0030
2.26	0.0310	0.9881	0.0119	2.76	0.0089	0.9971	0.0029
2.27	0.0303	0.9884	0.0116	2.77	0.0086	0.9972	0.0028
2.28	0.0296	0.9887	0.0113	2.78	0.0084	0.9973	0.0027
2.29	0.0290	0.9890	0.0110	2.79	0.0081	0.9974	0.0026
2.30	0.0283	0.9893	0.0107	2.80	0.0079	0.9974	0.0026
2.31	0.0277	0.9896	0.0104	2.81	0.0077	0.9975	0.0025
2.32	0.0271	0.9898	0.0102	2.82	0.0075	0.9976	0.0024
2.33	0.0264	0.9901	0.0099	2.83	0.0073	0.9977	0.0023
2.34	0.0258	0.9904	0.0096	2.84	0.0071	0.9977	0.0023
2.35	0.0252	0.9906	0.0094	2.85	0.0069	0.9978	0.0022
2.36	0.0246	0.9909	0.0091	2.86	0.0067	0.9979	0.0021
2.37	0.0241	0.9911	0.0089	2.87	0.0065	0.9980	0.0021
2.38	0.0235	0.9913	0.0087	2.88	0.0063	0.9980	0.0020
2.39	0.0229	0.9916	0.0084	2.89	0.0061	0.9981	0.0019
2.40	0.0224	0.9918	0.0082	2.90	0.0060	0.9981	0.0019
2.41	0.0219	0.9920	0.0080	2.91	0.0058	0.9982	0.0018
2.42	0.0213	0.9922	0.0078	2.92	0.0056	0.9982	0.0018
2.43	0.0208	0.9925	0.0076	2.93	0.0054	0.9983	0.0017
2.44	0.0203	0.9927	0.0073	2.94	0.0053	0.9984	0.0016
2.45	0.0198	0.9929	0.0071	2.95	0.0051	0.9984	0.0016
2.46	0.0194	0.9930	0.0069	2.96	0.0050	0.9985	0.0015
2.47	0.0189	0.9932	0.0068	2.97	0.0049	0.9985	0.0015
2.48	0.0184	0.9934	0.0066	2.98	0.0047	0.9986	0.0014
2.49	0.0180	0.9936	0.0064	2.99	0.0046	0.9986	0.0014
2.50	0.0175	0.9938	0.0062	3.00	0.0044	0.9987	0.0014

Normal distribution

z	$f(z)$	$F(z)$	$1-F(z)$	z	$f(z)$	$F(z)$	$1-F(z)$
3.00	0.0044	0.9987	0.0014	3.50	0.0009	0.9998	0.0002
3.01	0.0043	0.9987	0.0013	3.51	0.0008	0.9998	0.0002
3.02	0.0042	0.9987	0.0013	3.52	0.0008	0.9998	0.0002
3.03	0.0040	0.9988	0.0012	3.53	0.0008	0.9998	0.0002
3.04	0.0039	0.9988	0.0012	3.54	0.0008	0.9998	0.0002
3.05	0.0038	0.9989	0.0011	3.55	0.0007	0.9998	0.0002
3.06	0.0037	0.9989	0.0011	3.56	0.0007	0.9998	0.0002
3.07	0.0036	0.9989	0.0011	3.57	0.0007	0.9998	0.0002
3.08	0.0035	0.9990	0.0010	3.58	0.0007	0.9998	0.0002
3.09	0.0034	0.9990	0.0010	3.59	0.0006	0.9998	0.0002
3.10	0.0033	0.9990	0.0010	3.60	0.0006	0.9998	0.0002
3.11	0.0032	0.9991	0.0009	3.61	0.0006	0.9999	0.0001
3.12	0.0031	0.9991	0.0009	3.62	0.0006	0.9999	0.0001
3.13	0.0030	0.9991	0.0009	3.63	0.0006	0.9999	0.0001
3.14	0.0029	0.9992	0.0008	3.64	0.0005	0.9999	0.0001
3.15	0.0028	0.9992	0.0008	3.65	0.0005	0.9999	0.0001
3.16	0.0027	0.9992	0.0008	3.66	0.0005	0.9999	0.0001
3.17	0.0026	0.9992	0.0008	3.67	0.0005	0.9999	0.0001
3.18	0.0025	0.9993	0.0007	3.68	0.0005	0.9999	0.0001
3.19	0.0025	0.9993	0.0007	3.69	0.0004	0.9999	0.0001
3.20	0.0024	0.9993	0.0007	3.70	0.0004	0.9999	0.0001
3.21	0.0023	0.9993	0.0007	3.71	0.0004	0.9999	0.0001
3.22	0.0022	0.9994	0.0006	3.72	0.0004	0.9999	0.0001
3.23	0.0022	0.9994	0.0006	3.73	0.0004	0.9999	0.0001
3.24	0.0021	0.9994	0.0006	3.74	0.0004	0.9999	0.0001
3.25	0.0020	0.9994	0.0006	3.75	0.0003	0.9999	0.0001
3.26	0.0020	0.9994	0.0006	3.76	0.0003	0.9999	0.0001
3.27	0.0019	0.9995	0.0005	3.77	0.0003	0.9999	0.0001
3.28	0.0018	0.9995	0.0005	3.78	0.0003	0.9999	0.0001
3.29	0.0018	0.9995	0.0005	3.79	0.0003	0.9999	0.0001
3.30	0.0017	0.9995	0.0005	3.80	0.0003	0.9999	0.0001
3.31	0.0017	0.9995	0.0005	3.81	0.0003	0.9999	0.0001
3.32	0.0016	0.9996	0.0004	3.82	0.0003	0.9999	0.0001
3.33	0.0016	0.9996	0.0004	3.83	0.0003	0.9999	0.0001
3.34	0.0015	0.9996	0.0004	3.84	0.0003	0.9999	0.0001
3.35	0.0015	0.9996	0.0004	3.85	0.0002	0.9999	0.0001
3.36	0.0014	0.9996	0.0004	3.86	0.0002	0.9999	0.0001
3.37	0.0014	0.9996	0.0004	3.87	0.0002	1.0000	0.0001
3.38	0.0013	0.9996	0.0004	3.88	0.0002	1.0000	0.0001
3.39	0.0013	0.9997	0.0003	3.89	0.0002	1.0000	0.0001
3.40	0.0012	0.9997	0.0003	3.90	0.0002	1.0000	0.0001
3.41	0.0012	0.9997	0.0003	3.91	0.0002	1.0000	0.0001
3.42	0.0011	0.9997	0.0003	3.92	0.0002	1.0000	0.0000
3.43	0.0011	0.9997	0.0003	3.93	0.0002	1.0000	0.0000
3.44	0.0011	0.9997	0.0003	3.94	0.0002	1.0000	0.0000
3.45	0.0010	0.9997	0.0003	3.95	0.0002	1.0000	0.0000
3.46	0.0010	0.9997	0.0003	3.96	0.0002	1.0000	0.0000
3.47	0.0010	0.9997	0.0003	3.97	0.0001	1.0000	0.0000
3.48	0.0009	0.9998	0.0003	3.98	0.0001	1.0000	0.0000
3.49	0.0009	0.9998	0.0002	3.99	0.0001	1.0000	0.0000
3.50	0.0009	0.9998	0.0002	4.00	0.0001	1.0000	0.0000

7.2 CRITICAL VALUES

Table 7.1 lists common critical values for a standard normal random variable, z_α, defined by (see Figure 7.2):

$$\text{Prob}[Z \geq z_\alpha] = \alpha. \tag{7.3}$$

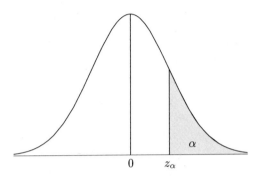

Figure 7.2: Critical values for a normal random variable.

α	z_α	α	z_α	α	z_α
.10	1.2816	.00009	3.7455	.000001	4.75
.05	1.6449	.00008	3.7750	.0000001	5.20
.025	1.9600	.00007	3.8082	.00000001	5.61
.01	2.3263	.00006	3.8461	.000000001	6.00
.005	2.5758	.00005	3.8906	.0000000001	6.36
.0025	2.8070	.00004	3.9444		
.001	3.0902	.00003	4.0128		
.0005	3.2905	.00002	4.1075		
.0001	3.7190	.00001	4.2649		

Table 7.1: Common critical values.

7.3 TOLERANCE FACTORS FOR NORMAL DISTRIBUTIONS

Suppose $X_1, X_2, \ldots, X_n$ is a random sample of size n from a normal population with mean μ and standard deviation σ. Using the summary statistics $\bar{x}$ and s, a tolerance interval $[L, U]$ may be constructed to capture $100P\%$ of the population with probability $1 - \alpha$. The following procedures may be used.

(1) Two-sided tolerance interval: A $100(1 - \alpha)\%$ tolerance interval that captures $100P\%$ of the population has as endpoints

$$[L, U] = \bar{x} \pm K_{\alpha, n, P} \cdot s \tag{7.4}$$

(2) One-sided tolerance interval, upper tailed: A $100(1-\alpha)\%$ tolerance interval bounded below has

$$L = \bar{x} - k_{\alpha,n,P} \cdot s \qquad U = \infty \qquad (7.5)$$

(3) One-sided tolerance interval, lower tailed: A $100(1-\alpha)\%$ tolerance interval bounded above has

$$L = -\infty \qquad U = \bar{x} + k_{\alpha,n,P} \cdot s \qquad (7.6)$$

where $K_{\alpha,n,P}$ is the tolerance factor given in section 7.3.1 and $k_{\alpha,n,P}$ is computed using the formula below.

Values of $K_{\alpha,n,P}$ are given in section 7.3.1 for $P = 0.75, 0.90, 0.95, 0.99, 0.999$, $\alpha = 0.75, 0.90, 0.95, 0.99$, and various values of n. The value of $k_{\alpha,n,P}$ is given by

$$k_{\alpha,n,P} = \frac{z_{1-P} + \sqrt{(z_{1-P})^2 - ab}}{a}$$

$$a = 1 - \frac{(z_\alpha)^2}{2(n-1)} \qquad (7.7)$$

$$b = z_{1-P}^2 - \frac{z_\alpha^2}{n}$$

where z_{1-P} and z_α are critical values for a standard normal random variable (see page 125).

Example 7.40: Suppose a sample of size $n = 30$ from a normal distribution has $\bar{x} = 10.02$ and $s = 0.13$. Find tolerance intervals with a confidence level 95% ($\alpha = .05$) and $P = .90$.

Solution:

(S1) Two-sided interval:
1. From the tables in section 7.3.1 we find $K_{.05,30,.90} = 2.413$.
2. The interval is $\bar{x} \pm K \cdot s = 10.02 \pm 0.31$; or $I = [9.71, 10.33]$.
3. We conclude: in each sample of size 30, at least 90% of the normal population being sampled will be in the interval I, with probability 95%.

(S2) One-sided intervals:
1. The critical values used in equation (7.7) are $z_{1-P} = z_{.10} = 1.282$ and $z_\alpha = z_{.05} = 1.645$. Using this equation: $a = 1 - \frac{(1.645)^2}{2(29)} = 0.9533$, $b = (1.282)^2 - \frac{(1.645)^2}{30} = 1.553$, and $k_{.05,30,.90} = 1.768$
2. The lower bound is $L = \bar{x} - k \cdot s = 9.79$.
3. The upper bound is $U = \bar{x} + k \cdot s = 10.25$.
4. We conclude:
   (a) In each sample of size 30, at least 90% of the normal population being sampled will be greater than L, with probability 95%.
   (b) In each sample of size 30, at least 90% of the normal population being sampled will be smaller than U, with probability 95%.

7.3.1 Tables of tolerance intervals for normal distributions

Tolerance factors for normal distributions

		$P = .90$							
n	$\alpha = .10$	.05	.01	.001	n	$\alpha = .10$	.05	.01	.001
2	15.978	18.800	24.167	30.227	20	2.152	2.564	3.368	4.300
3	5.847	6.919	8.974	11.309	25	2.077	2.474	3.251	4.151
4	4.166	4.943	6.440	8.149	30	2.025	2.413	3.170	4.049
5	3.494	4.152	5.423	6.879	40	1.959	2.334	3.066	3.917
6	3.131	3.723	4.870	6.188	50	1.916	2.284	3.001	3.833
7	2.902	3.452	4.521	5.750	75	1.856	2.211	2.906	3.712
8	2.743	3.264	4.278	5.446	100	1.822	2.172	2.854	3.646
9	2.626	3.125	4.098	5.220	500	1.717	2.046	2.689	3.434
10	2.535	3.018	3.959	5.046	1000	1.695	2.019	2.654	3.390
15	2.278	2.713	3.562	4.545	∞	1.645	1.960	2.576	3.291

Tolerance factors for normal distributions

		$P = .95$							
n	$\alpha = .10$	.05	.01	.001	n	$\alpha = .10$	.05	.01	.001
2	32.019	37.674	48.430	60.573	20	2.310	2.752	3.615	4.614
3	8.380	9.916	12.861	16.208	25	2.208	2.631	3.457	4.413
4	5.369	6.370	8.299	10.502	30	2.140	2.549	3.350	4.278
5	4.275	5.079	6.634	8.415	40	2.052	2.445	3.213	4.104
6	3.712	4.414	5.775	7.337	50	1.996	2.379	3.126	3.993
7	3.369	4.007	5.248	6.676	75	1.917	2.285	3.002	3.835
8	3.136	3.732	4.891	6.226	100	1.874	2.233	2.934	3.748
9	2.967	3.532	4.631	5.899	500	1.737	2.070	2.721	3.475
10	2.839	3.379	4.433	5.649	1000	1.709	2.036	2.676	3.418
15	2.480	2.954	3.878	4.949	∞	1.645	1.960	2.576	3.291

Tolerance factors for normal distributions

		$P = .99$							
n	$\alpha = .10$	.05	.01	.001	n	$\alpha = .10$	.05	.01	.001
2	160.193	188.491	242.300	303.054	20	2.659	3.168	4.161	5.312
3	18.930	22.401	29.055	36.616	25	2.494	2.972	3.904	4.985
4	9.398	11.150	14.527	18.383	30	2.385	2.841	3.733	4.768
5	6.612	7.855	10.260	13.015	40	2.247	2.677	3.518	4.493
6	5.337	6.345	8.301	10.548	50	2.162	2.576	3.385	4.323
7	4.613	5.488	7.187	9.142	75	2.042	2.433	3.197	4.084
8	4.147	4.936	6.468	8.234	100	1.977	2.355	3.096	3.954
9	3.822	4.550	5.966	7.600	500	1.777	2.117	2.783	3.555
10	3.582	4.265	5.594	7.129	1000	1.736	2.068	2.718	3.472
15	2.945	3.507	4.605	5.876	∞	1.645	1.960	2.576	3.291

7.4 OPERATING CHARACTERISTIC CURVES

7.4.1 One-sample Z test

Consider a one-sample hypothesis test on a population mean of a normal distribution with known standard deviation σ (see section 10.2). The general form of the hypothesis test (for each possible alternative hypothesis) is:

$H_0: \mu = \mu_0$

$H_a: \mu > \mu_0, \quad \mu < \mu_0, \quad \mu \neq \mu_0$

TS: $Z = \dfrac{\bar{X} - \mu_0}{\sigma/\sqrt{n}}$

RR: $Z \geq z_\alpha, \quad Z \leq -z_\alpha, \quad |Z| \geq z_{\alpha/2}$

Let α be the probability of a Type I error, β the probability of a Type II error, and μ_a an alternative mean. For $\Delta = |\mu_a - \mu_0|/\sigma$ the operating characteristic curve returns the probability of not rejecting the null hypothesis given $\mu = \mu_a$. The curves may be used to determine the appropriate sample size for given values of α, β, and Δ.

7.4.2 Two-sample Z test

Consider a two-sample hypothesis test for comparing population means from normal distributions with known standard deviations σ_1 and σ_2 (see section 10.3). The general form of the hypothesis test for testing the equality of means (for each possible alternative hypothesis) is:

$H_0: \mu_1 - \mu_2 = 0$

$H_a: \mu_1 - \mu_2 > 0, \quad \mu_1 - \mu_2 < 0, \quad \mu_1 - \mu_2 \neq 0$

TS: $Z = \dfrac{\bar{X}_1 - \bar{X}_2}{\sqrt{\dfrac{\sigma_1^2}{n_1} + \dfrac{\sigma_2^2}{n_2}}}$

RR: $Z \geq z_\alpha, \quad Z \leq -z_\alpha, \quad |Z| \geq z_{\alpha/2}$

Let α be the probability of a Type I error and β the probability of a Type II error. For given values of α, and $\Delta = \dfrac{|\mu_1 - \mu_2|}{\sqrt{\sigma_1^2 + \sigma_2^2}}$ the operating characteristic curve returns the probability of not rejecting the null hypothesis. The curves may be used to determine an appropriate sample size ($n = n_1 = n_2$) for desired levels of α, β, and Δ.

7.4. OPERATING CHARACTERISTIC CURVES

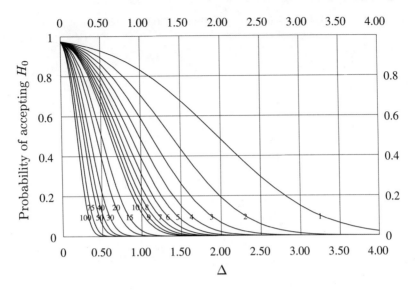

Figure 7.3: Operating characteristic curves, various values of n, Z test, two-sided alternative, $\alpha = .05$.

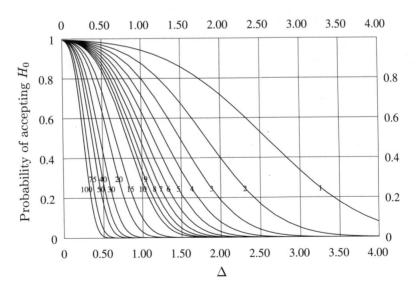

Figure 7.4: Operating characteristic curves, various values of n, Z test, two-sided alternative, $\alpha = .01$.

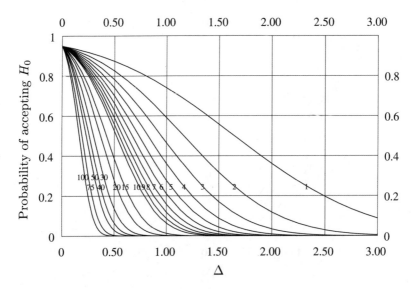

Figure 7.5: Operating characteristic curves, various values of n, Z test, one-sided alternative, $\alpha = .05$.

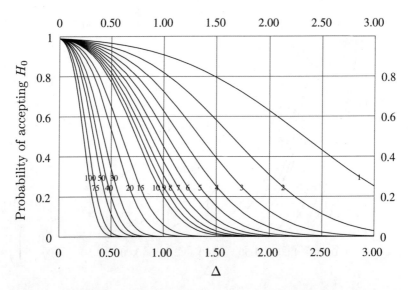

Figure 7.6: Operating characteristic curves, various values of n, Z test, one-sided alternative, $\alpha = .01$.

7.5 MULTIVARIATE NORMAL DISTRIBUTION

Let each $\{X_i\}$ (for $i = 1,\ldots,n$) be a normal random variable with mean μ_i and variance σ_{ii}. If the covariance of X_i and X_j is σ_{ij}, then the joint probability density of the $\{X_i\}$ is:

$$f(\mathbf{x}) = \frac{1}{(2\pi)^{n/2}\sqrt{\det(\boldsymbol{\Sigma})}} \exp\left[-\frac{1}{2}(\mathbf{x}-\boldsymbol{\mu})^{\mathrm{T}}\boldsymbol{\Sigma}^{-1}(\mathbf{x}-\boldsymbol{\mu})\right] \quad (7.8)$$

where

(a) $\mathbf{x} = \begin{bmatrix} x_1 & x_2 & \ldots & x_n \end{bmatrix}^{\mathrm{T}}$

(b) $\boldsymbol{\mu} = \begin{bmatrix} \mu_1 & \mu_2 & \ldots & \mu_n \end{bmatrix}^{\mathrm{T}}$

(c) $\boldsymbol{\Sigma}$ is an $n \times n$ matrix with elements σ_{ij}

The corresponding characteristic function is

$$\phi(\mathbf{t}) = \exp\left[i\boldsymbol{\mu}^{\mathrm{T}}\mathbf{t} - \frac{1}{2}\mathbf{t}^{\mathrm{T}}\boldsymbol{\Sigma}\mathbf{t}\right] \quad (7.9)$$

The form of the characteristic function implies that all cumulants of higher order than 2 vanish (see Marcienkiewicz's theorem). Therefore, all moments of order higher than 2 may be expressed in terms of those of order 1 and 2. If $\boldsymbol{\mu} = \mathbf{0}$ then the odd moments vanish and the $(2n)^{\mathrm{th}}$ moment satisfies

$$\mathrm{E}\left[\underbrace{X_i X_j X_k X_l \cdots}_{2n \text{ terms}}\right] = \frac{(2n)!}{n!2^n}\{\sigma_{ij}\sigma_{kl}\cdots\}_{\mathrm{sym}} \quad (7.10)$$

where the subscript "sym" means the symmetrized form of the product of the σ's.

See C. W. Gardiner *Handbook of Stochastic Methods*, Springer-Verlag, New York, 1985, pages 36–37.

7.6 DISTRIBUTION OF THE CORRELATION COEFFICIENT FOR A BIVARIATE NORMAL

The bivariate normal probability function is given by

$$f(x,y) = \frac{1}{2\pi\sigma_x\sigma_y\sqrt{1-\rho^2}} \exp\left(-\frac{1}{2(1-\rho^2)}\right. \quad (7.11)$$

$$\left. \times \left[\left(\frac{x-\mu_x}{\sigma_x}\right)^2 - 2\rho\left(\frac{x-\mu_x}{\sigma_x}\right)\left(\frac{y-\mu_y}{\sigma_y}\right) + \left(\frac{y-\mu_y}{\sigma_y}\right)^2\right]\right)$$

where μ_x = mean of x
μ_y = mean of y
σ_x = standard deviation of x
σ_y = standard deviation of y
ρ = correlation coefficient between x and y

Given a sample $\{(x_1, y_1), \ldots, (x_n, y_n)\}$ of size n, the sample correlation coefficient, an estimate of ρ, is

$$r = \frac{\sum_{i=1}^{n} (x_i - \bar{x})(y_i - \bar{y})}{\sqrt{\left(\sum_{i=1}^{n} (x_i - \bar{x})^2\right)\left(\sum_{i=1}^{n} (y_i - \bar{y})^2\right)}} \quad (7.12)$$

where $\bar{x} = \left(\sum_{i=1}^{n} x_i\right)/n$ and $\bar{y} = \left(\sum_{i=1}^{n} y_i\right)/n$. The moments are given by:

$$\begin{aligned}
\mu_r &= \rho - \frac{\rho(1-\rho^2)}{(n+1)} \\
\sigma_r^2 &= \frac{(1-\rho^2)^2}{n+1}\left(1 + \frac{11\rho^2}{2(n+1)} + \cdots\right) \\
\gamma_1 &= \frac{6\rho}{\sqrt{n+1}}\left(1 + \frac{77\rho^2 - 30}{12(n+1)} + \cdots\right) \\
\gamma_2 &= \frac{6}{n+1}\left(12\rho^2 - 1\right) + \cdots
\end{aligned} \quad (7.13)$$

7.6.1 Normal approximation

If r is the sample correlation coefficient (defined in equation (7.12)), the random variable

$$Z = \tanh^{-1} r = \frac{1}{2} \ln \frac{1+r}{1-r} \quad (7.14)$$

is approximately normally distributed with parameters

$$\mu_Z = \frac{1}{2} \ln \left[\frac{1+\rho}{1-\rho}\right] = \tanh^{-1} \rho \quad \text{and} \quad \sigma_Z^2 = \frac{1}{n-3} \quad (7.15)$$

7.6.2 Zero correlation coefficient for bivariate normal

In the special case where $\rho = 0$, the density function of r becomes

$$f_n(r; 0) = \frac{1}{\sqrt{\pi}} \frac{\Gamma((n-1)/2)}{\Gamma((n-2)/2)} \left(1 - r^2\right)^{(n-4)/2} \quad (7.16)$$

Under the transformation

$$r^2 = \frac{t^2}{t^2 + \nu} \quad (7.17)$$

7.6. DISTRIBUTION OF THE CORRELATION COEFFICIENT

$f_n(r; 0)$, as given by equation (7.16), has a t-distribution with $\nu = n-1$ degrees of freedom. The following table gives percentage points of the distribution of the correlation coefficient when $\rho = 0$.

Percentage points of the correlation coefficient, when $\rho = 0$
Prob $[r \leq$ tabulated value$] = 1 - \alpha$

$\alpha =$	0.05	0.025	0.01	0.005	0.0025	0.0005
$2\alpha =$	0.1	0.05	0.02	0.01	0.005	0.001
$\nu = 1$	0.988	0.997	$0.9^3 507$	$0.9^3 877$	$0.9^4 69$	0.9^6
2	0.900	0.950	0.980	0.990	0.995	0.999
3	0.805	0.878	0.934	0.959	0.974	0.991
4	0.729	0.811	0.882	0.917	0.942	0.974
5	0.669	0.754	0.833	0.875	0.906	0.951
6	0.621	0.707	0.789	0.834	0.870	0.925
7	0.582	0.666	0.750	0.798	0.836	0.898
8	0.549	0.632	0.715	0.765	0.805	0.872
9	0.521	0.602	0.685	0.735	0.776	0.847
10	0.497	0.576	0.658	0.708	0.750	0.823
11	0.476	0.553	0.634	0.684	0.726	0.801
12	0.458	0.532	0.612	0.661	0.703	0.780
13	0.441	0.514	0.592	0.641	0.683	0.760
14	0.426	0.497	0.574	0.623	0.664	0.742
15	0.412	0.482	0.558	0.606	0.647	0.725
16	0.400	0.468	0.543	0.590	0.631	0.708
17	0.389	0.456	0.529	0.575	0.616	0.693
18	0.378	0.444	0.516	0.561	0.602	0.679
19	0.369	0.433	0.503	0.549	0.589	0.665
20	0.360	0.423	0.492	0.537	0.576	0.652
25	0.323	0.381	0.445	0.487	0.524	0.597
30	0.296	0.349	0.409	0.449	0.484	0.554
35	0.275	0.325	0.381	0.418	0.452	0.519
40	0.257	0.304	0.358	0.393	0.425	0.490
45	0.243	0.288	0.338	0.372	0.403	0.465
50	0.231	0.273	0.322	0.354	0.384	0.443
60	0.211	0.250	0.295	0.325	0.352	0.408
70	0.195	0.232	0.274	0.302	0.327	0.380
80	0.183	0.217	0.257	0.283	0.307	0.357
90	0.173	0.205	0.242	0.267	0.290	0.338
100	0.164	0.195	0.230	0.254	0.276	0.321

Use the α value for a single-tail test. For a two-tail test, use the 2α value. If r is computed from n paired observations, enter the table with $\nu = n - 2$. For partial correlations, enter the table with $\nu = n - 2 - k$, where k is the number of variables held constant.

CHAPTER 8

Estimation

A nonconstant function of a set of random variables is a *statistic*. It is a function of observable random variables, which does not contain any unknown parameters. A statistic is itself an observable random variable.

Let θ be a parameter appearing in the density function for the random variable X. Let g be a function that returns an approximate value $\hat{\theta}$ of θ from a given sample $\{x_1, \ldots, x_n\}$. Then $\hat{\theta} = g(x_1, x_2, \ldots, x_n)$ may be considered a single observation of the random variable $\hat{\Theta} = g(X_1, X_2, \ldots, X_n)$. The random variable $\hat{\Theta}$ is an *estimator* for the parameter θ.

8.1 DEFINITIONS

(1) $\hat{\Theta}$ is an **unbiased estimator** for θ if $\mathrm{E}\,[\hat{\Theta}] = \theta$.

(2) The **bias** of the estimator $\hat{\Theta}$ is $\mathrm{B}\,[\hat{\Theta}] = \mathrm{E}\,[\hat{\Theta}] - \theta$.

(3) The **mean square error** of $\hat{\Theta}$ is
$$\mathrm{MSE}\,[\hat{\Theta}] = \mathrm{E}\,[(\hat{\Theta} - \theta)^2] = \mathrm{Var}\,[\hat{\Theta}] + \mathrm{B}\,[\hat{\Theta}]^2\,.$$

(4) The **error of estimation** is $\epsilon = |\hat{\Theta} - \theta|$.

(5) Let $\hat{\Theta}_1$ and $\hat{\Theta}_2$ be unbiased estimators for θ.

   (a) If $\mathrm{Var}\,[\hat{\Theta}_1] < \mathrm{Var}\,[\hat{\Theta}_2]$ then the estimator $\hat{\Theta}_1$ is **relatively more efficient** than the estimator $\hat{\Theta}_2$.

   (b) The **efficiency of $\hat{\Theta}_2$ relative to $\hat{\Theta}_1$** is
   $$\mathrm{Efficiency} = \frac{\mathrm{Var}\,[\hat{\Theta}_1]}{\mathrm{Var}\,[\hat{\Theta}_2]}.$$

(6) $\hat{\Theta}$ is a **consistent estimator** for θ if for every $\epsilon > 0$,
$$\lim_{n \to \infty} \mathrm{Prob}\,[|\hat{\Theta} - \theta| \le \epsilon] = 1 \quad \text{or, equivalently}$$
$$\lim_{n \to \infty} \mathrm{Prob}\,[|\hat{\Theta} - \theta| > \epsilon] = 0.$$

(7) $\hat{\Theta}$ is a **sufficient estimator** for θ if for each value of $\hat{\Theta}$ the conditional distribution of $X_1, X_2, \ldots, X_n$ given $\hat{\Theta} = \theta_0$ is independent of θ.

(8) Let $\widehat{\Theta}$ be an estimator for the parameter θ and suppose $\widehat{\Theta}$ has sampling distribution $g(\widehat{\Theta})$. Then $\widehat{\Theta}$ is a **complete** statistic if for all θ, $\mathrm{E}\left[h\left(\widehat{\Theta}\right)\right] = 0$ implies $h\left(\widehat{\Theta}\right) = 0$ for all functions $h\left(\widehat{\Theta}\right)$.

(9) An estimator $\widehat{\Theta}$ for θ is a **minimum variance unbiased estimator** (MVUE) if it is unbiased and has the smallest possible variance.

Example 8.41: To determine if $\overline{X}^2$ is an unbiased estimator of μ^2, consider the following expected value:

$$\begin{aligned}
\mathrm{E}\left[\overline{X}^2\right] &= \mathrm{E}\left[\left(\frac{1}{n}\sum_{i=1}^n X_i\right)^2\right] \\
&= \frac{1}{n^2}\mathrm{E}\left[\sum_{i=1}^n X_i^2 + \sum_{i=1,j=1,i\neq j}^n X_i X_j\right] \\
&= \frac{1}{n^2}\left[n(\sigma^2 + \mu^2) + (n^2 - n)\mu^2\right] \\
&= \mu^2 + \frac{\sigma^2}{n} \\
&> \mu^2
\end{aligned} \qquad (8.1)$$

This shows $\overline{X}^2$ is a biased estimator of μ^2.

8.2 CRAMÉR–RAO INEQUALITY

Let $\{X_1, X_2, \ldots, X_n\}$ be a random sample from a population with probability density function $f(x)$. Let $\widehat{\Theta}$ be an unbiased estimator for θ. Under very general conditions it can be shown that

$$\mathrm{Var}\left[\widehat{\Theta}\right] \geq \frac{1}{n \cdot \mathrm{E}\left[\left(\frac{\partial \ln f(X)}{\partial \theta}\right)^2\right]} = \frac{1}{-n \cdot \mathrm{E}\left[\frac{\partial^2 \ln f(X)}{\partial \theta^2}\right]}. \qquad (8.2)$$

If equality holds then $\widehat{\Theta}$ is a minimum variance unbiased estimator (MVUE) for θ.

Example 8.42: The probability density function for a normal random variable with unknown mean θ and known variance σ^2 is $f(x;\theta) = \frac{1}{\sqrt{2\pi}\sigma}\exp\left[-\frac{(x-\theta)^2}{2\sigma^2}\right]$. Use the Cramér–Rao inequality to show the minimum variance of any unbiased estimator, $\widehat{\Theta}$, for θ is at least σ^2/n.

Solution:

(S1) $\dfrac{\partial}{\partial \theta}\ln f(X;\theta) = \dfrac{(X-\theta)}{\sigma^2}$

(S2) $\left(\dfrac{\partial \ln f(X;\theta)}{\partial \theta}\right)^2 = \dfrac{(X-\theta)^2}{\sigma^4}$

(S3) $E\left[\dfrac{(X-\theta)^2}{\sigma^4}\right] = \displaystyle\int_{-\infty}^{\infty} \dfrac{(x-\theta)^2}{\sigma^4}\left(\dfrac{1}{\sqrt{2\pi}\sigma}e^{-(x-\theta)^2/2\sigma^2}\right)dx = \dfrac{1}{\sigma^2}$

(S4) $\operatorname{Var}[\widehat{\Theta}] \geq \dfrac{1}{n\frac{1}{\sigma^2}} = \dfrac{\sigma^2}{n}$

8.3 THEOREMS

(1) $\widehat{\Theta}$ is a consistent estimator for θ if
  (a) $\widehat{\Theta}$ is unbiased, and
  (b) $\displaystyle\lim_{n\to\infty} \operatorname{Var}[\widehat{\Theta}] = 0$.

(2) $\widehat{\Theta}$ is a sufficient estimator for the parameter θ if the joint distribution of $\{X_1, X_2, \ldots, X_n\}$ can be factored into

$$f(x_1, x_2, \ldots, x_n; \theta) = g(\widehat{\Theta}, \theta) \cdot h(x_1, x_2, \ldots, x_n) \qquad (8.3)$$

where $g(\widehat{\Theta}, \theta)$ depends only on the estimate $\widehat{\theta}$ and the parameter θ, and $h(x_1, x_2, \ldots, x_n)$ does not depend on the parameter θ.

(3) Unbiased estimators:
  (a) An unbiased estimator may not exist.
  (b) An unbiased estimator is not unique.
  (c) An unbiased estimator may be meaningless.
  (d) An unbiased estimator is not necessarily consistent.

(4) Consistent estimators:
  (a) A consistent estimator is not unique.
  (b) A consistent estimator may be meaningless.
  (c) A consistent estimator is not necessarily unbiased.

(5) Maximum likelihood estimators (MLE):
  (a) A MLE need not be consistent.
  (b) A MLE may not be unbiased.
  (c) A MLE is not unique.
  (d) If a single sufficient statistic T exists for the parameter θ, the MLE of θ must be a function of T.
  (e) Let $\widehat{\Theta}$ be a MLE of θ. If $\tau(\cdot)$ is a function with a single-valued inverse, then a MLE of $\tau(\theta)$ is $\tau(\widehat{\Theta})$.

(6) Method of moments (MOM) estimators:
  (a) MOM estimators are not uniquely defined.
  (b) MOM estimators may not be functions of sufficient or complete statistics.

(7) A single sufficient estimator may not exist.

8.4 THE METHOD OF MOMENTS

The moment estimators are the solutions to the systems of equations

$$\mu'_r = \mathrm{E}[X^r] = \frac{1}{n}\sum_{i=1}^{n} x_i^r = m'_r, \quad r = 1, 2, \ldots, k$$

where k is the number of parameters.

8.5 THE LIKELIHOOD FUNCTION

Let $x_1, x_2, \ldots, x_n$ be the values of a random sample from a population characterized by the parameters $\theta_1, \theta_2, \ldots, \theta_r$. The **likelihood function** of the sample is

(1) the joint probability mass function evaluated at $(x_1, x_2, \ldots, x_n)$ if $(X_1, X_2, \ldots, X_n)$ are discrete,

$$\mathrm{L}(\theta_1, \theta_2, \ldots, \theta_r) = p(x_1, x_2, \ldots, x_n; \theta_1, \theta_2, \ldots, \theta_r) \quad (8.4)$$

(2) the joint probability density function evaluated at $(x_1, x_2, \ldots, x_n)$ if $(X_1, X_2, \ldots, X_n)$ are continuous.

$$\mathrm{L}(\theta_1, \theta_2, \ldots, \theta_r) = f(x_1, x_2, \ldots, x_n; \theta_1, \theta_2, \ldots, \theta_r) \quad (8.5)$$

8.6 THE METHOD OF MAXIMUM LIKELIHOOD

The maximum likelihood estimators (MLEs) are those values of the parameters that maximize the likelihood function of the sample: $\mathrm{L}(\theta_1, \ldots, \theta_r)$. In practice, it is often easier to maximize $\ln \mathrm{L}(\theta_1, \ldots, \theta_r)$. This is equivalent to maximizing the likelihood function, $\mathrm{L}(\theta_1, \ldots, \theta_r)$, since $\ln \mathrm{L}(\theta_1, \ldots, \theta_r)$ is a monotonic function of $\mathrm{L}(\theta_1, \ldots, \theta_r)$.

Example 8.43: Suppose $X_1, X_2, \ldots, X_n$ is a random sample from a population having a Poisson distribution with parameter λ. Find the maximum likelihood estimator for the parameter λ.

Solution:

(S1) The probability mass function for a Poisson random variable is
$$f(x; \lambda) = \frac{e^{-\lambda} \lambda^x}{x!}$$

(S2) We compute

$$\mathrm{L}(\theta) = \left(\frac{e^{-\lambda}\lambda^{x_1}}{x_1!}\right)\left(\frac{e^{-\lambda}\lambda^{x_2}}{x_2!}\right)\cdots\left(\frac{e^{-\lambda}\lambda^{x_n}}{x_n!}\right)$$

$$= \frac{e^{-n\lambda}\lambda^{x_1+x_2+\cdots+x_n}}{x_1! x_2! \cdots x_n!} \quad (8.6)$$

$$\ln \mathrm{L}(\theta) = -n\lambda + (x_1 + x_2 + \cdots + x_n)\ln\lambda + \ln(x_1! x_2! \cdots x_n!)$$

(S3) $\dfrac{\partial \ln \mathrm{L}(\lambda)}{\partial \lambda} = -n + \dfrac{x_1 + x_2 + \cdots + x_n}{\lambda} = 0$

8.7. INVARIANCE PROPERTY OF MLES

(S4) Solving for λ: $\widehat{\lambda} = \dfrac{x_1 + x_2 + \cdots + x_n}{n} = \overline{x}$ is the MLE for λ.

8.7 INVARIANCE PROPERTY OF MAXIMUM LIKELIHOOD ESTIMATORS

Let $\widehat{\Theta}_1, \widehat{\Theta}_2, \ldots, \widehat{\Theta}_r$ be the maximum likelihood estimators for $\theta_1, \theta_2, \ldots, \theta_r$ and let $h(\theta_1, \theta_2, \ldots, \theta_r)$ be a function of $\theta_1, \theta_2, \ldots, \theta_r$. The maximum likelihood estimator of the parameter $h(\theta_1, \theta_2, \ldots, \theta_r)$ is

$$\widehat{h}(\theta_1, \theta_2, \ldots, \theta_r) = h(\widehat{\Theta}_1, \widehat{\Theta}_2, \ldots, \widehat{\Theta}_r).$$

8.8 DIFFERENT ESTIMATORS

Assume $\{x_1, x_2, \ldots, x_n\}$ is a set of observations. Let UMV stand for *uniformly minimum variance unbiased* and let MLE stand for *maximum likelihood estimator*.

(1) Normal distribution: $N(\mu, \sigma^2)$
   (a) When σ is known:
      1. $\sum x_i$ is necessary, sufficient, and complete.
      2. Point estimate for μ: $\widehat{\mu} = \overline{x} = \dfrac{1}{n}\sum x_i$ is UMV, MLE.
   (b) When μ is known:
      1. $\sum (x_i - \mu)^2$ is necessary, sufficient, and complete.
      2. Point estimate for σ^2: $\widehat{\sigma^2} = \dfrac{\sum (x_i - \mu)^2}{n}$ is UMV, MLE.
   (c) When μ and σ are unknown:
      1. $\{\sum x_i, \sum (x_i - \overline{x})^2\}$ are necessary, sufficient, and complete.
      2. Point estimate for μ: $\widehat{\mu} = \overline{x} = \dfrac{1}{n}\sum x_i$ is UMV, MLE.
      3. Point estimate for σ^2: $\widehat{\sigma^2} = \dfrac{\sum (x_i - \overline{x})^2}{n}$ is MLE.
      4. Point estimate for σ^2: $\widehat{\sigma^2} = \dfrac{\sum (x_i - \overline{x})^2}{n-1}$ is UMV.
      5. Point estimate for σ: $\widehat{\sigma} = \dfrac{\Gamma[(n-1)/2]}{\sqrt{2}\,\Gamma(n/2)}\sqrt{\dfrac{\sum (x_i - \overline{x})^2}{n-1}}$ is UMV.

(2) Poisson distribution with parameter λ:
   (a) $\sum x_i$ is necessary, sufficient, and complete.
   (b) Point estimate for λ: $\widehat{\lambda} = \dfrac{1}{n}\sum x_i$ is UMV, MLE.

(3) Uniform distribution on an interval:
   (a) Interval is $[0, \theta]$
      1. $\max(x_i)$ is necessary, sufficient, and complete.
      2. Point estimate for θ: $\widehat{\Theta} = \max(x_i)$ is MLE.
      3. Point estimate for θ: $\widehat{\Theta} = \dfrac{n+1}{n}\max(x_i)$ is UMV.

(b) Interval is $[\alpha, \beta]$
  1. $\{\min(x_i), \max(x_i)\}$ are necessary, sufficient, and complete.
  2. Point estimate for α: $\widehat{\alpha} = \dfrac{n\min(x_i) - \max(x_i)}{n-1}$ is UMV.
  3. Point estimate for α: $\widehat{\alpha} = \min(x_i)$ is MLE.
  4. Point estimate for $\frac{\alpha+\beta}{2}$: $\widehat{\dfrac{\alpha+\beta}{2}} = \dfrac{\min(x_i) + \max(x_i)}{2}$ is UMV.

(c) Interval is $\left[\theta - \frac{1}{2}, \theta + \frac{1}{2}\right]$
  1. $\{\min(x_i), \max(x_i)\}$ are necessary and sufficient.
  2. Point estimate for θ: $\widehat{\Theta} = \dfrac{\min(x_i) + \max(x_i)}{2}$ is MLE.

CHAPTER 9

Confidence Intervals

9.1 DEFINITIONS

A simple point estimate $\hat{\theta}$ of a parameter θ serves as a best guess for the value of θ, but conveys no sense of confidence in the estimate. A confidence interval I, based on $\hat{\theta}$, is used to make statements about θ when the sample size, the underlying distribution of θ, and the **confidence coefficient** $1 - \alpha$ are known. We make statements of the form:

$$\text{The probability that } \theta \text{ is in a specified interval is } 1 - \alpha. \quad (9.1)$$

To construct a confidence interval for a parameter θ, the confidence coefficient must be specified. The usual procedure is to specify the confidence coefficient $1 - \alpha$ and then determine the confidence interval. A typical value is $\alpha = 0.05$ (also written as $\alpha = 5\%$); so that $1 - \alpha = 0.95$ (or 95% *confidence*).

The confidence interval may be denoted $(\theta_{\text{low}}, \theta_{\text{high}})$. There are many ways to specify θ_{low} and θ_{high}, depending on the parameter θ and the underlying distribution. The bounds on the confidence interval are usually defined to satisfy

$$\text{The probability that } \theta < \theta_{\text{low}} \text{ is } \alpha/2: \text{ Prob}\,[\theta < \theta_{\text{low}}] = \alpha/2$$
$$\text{The probability that } \theta > \theta_{\text{high}} \text{ is } \alpha/2: \text{ Prob}\,[\theta > \theta_{\text{high}}] = \alpha/2 \quad (9.2)$$

so that $\text{Prob}\,[\theta_{\text{low}} \leq \theta \leq \theta_{\text{high}}] = 1 - \alpha$.

It is also possible to construct one-sided confidence intervals. For these,

(1) $\theta_{\text{low}} = -\infty$ and $\text{Prob}\,[\theta > \theta_{\text{high}}] = \alpha$ or $\text{Prob}\,[\theta \leq \theta_{\text{high}}] = 1 - \alpha$
    (one-sided, lower-tailed confidence interval), or

(2) $\theta_{\text{high}} = \infty$ and $\text{Prob}\,[\theta < \theta_{\text{low}}] = \alpha$ or $\text{Prob}\,[\theta \geq \theta_{\text{low}}] = 1 - \alpha$
    (one-sided, upper-tailed confidence interval).

9.2 COMMON CRITICAL VALUES

The formulas for common confidence intervals usually involve critical values from the normal distribution, the t distribution, or the chi-square distribution; see Tables 9.3 and 9.4. Table 9.1 contains common critical values needed for confidence intervals.

		α			
Distribution	0.10	0.05	0.01	0.001	0.0001
t distribution					
$t_{\alpha/2, 10}$	1.8125	2.2281	3.1693	4.5869	6.2111
$t_{\alpha/2, 100}$	1.6602	1.9840	2.6259	3.3905	4.0533
$t_{\alpha/2, 1000}$	1.6464	1.9623	2.5808	3.3003	3.9063
Normal distribution					
$z_{\alpha/2}$	1.6449	1.9600	2.5758	3.2905	3.8906
χ^2 distribution					
$\chi^2_{1-\alpha/2, 10}$	3.9403	3.2470	2.1559	1.2650	0.7660
$\chi^2_{\alpha/2, 10}$	18.3070	20.4832	25.1882	31.4198	37.3107
$\chi^2_{1-\alpha/2, 100}$	77.9295	74.2219	67.3276	59.8957	54.1129
$\chi^2_{\alpha/2, 100}$	124.3421	129.5612	140.1695	153.1670	164.6591
$\chi^2_{1-\alpha/2, 1000}$	927.5944	914.2572	888.5635	859.3615	835.3493
$\chi^2_{\alpha/2, 1000}$	1074.6790	1089.5310	1118.9480	1153.7380	1183.4920
	0.90	0.95	0.99	0.999	0.9999
			$1 - \alpha$		

Table 9.1: Common critical values used with confidence intervals.

9.3 SAMPLE SIZE CALCULATIONS

In order to construct a confidence interval of specified width, a priori parameter estimates and a bound on the error of estimation may be used to determine the necessary sample size. For a $100(1 - \alpha)\%$ confidence interval, let E = error of estimation (half the width of the confidence interval). Table 9.2 presents some common sample size calculations.

Example 9.44: A researcher would like to estimate the probability of a success, p, in a binomial experiment. How large a sample is necessary in order to estimate this proportion to within .05 with 99% confidence, i.e., find a value of n such that Prob $[|\hat{p} - p| \leq 0.05] \geq 0.99$.

Solution:

(S1) Since no a priori estimate of p is available, use $p = .5$. The bound on the error of estimation is $E = .05$ and $1 - \alpha = .99$.

(S2) From Table 9.2, $n = \dfrac{z^2_{.005} \cdot pq}{E^2} = \dfrac{(2.5758)(.5)(.5)}{.05^2} = 663.47$.

(S3) This formula produces a conservative value for the necessary sample size (since no a priori estimate of p is known). A sample size of at least 664 should be used.

9.4. SUMMARY OF COMMON CONFIDENCE INTERVALS

Parameter	Estimate	Sample size	
μ	$\bar{x}$	$n = \left(\dfrac{z_{\alpha/2} \cdot \sigma}{E}\right)^2$	(1)
p	$\hat{p}$	$n = \dfrac{(z_{\alpha/2})^2 \cdot pq}{E^2}$	(2)
$\mu_2 - \mu_2$	$\bar{x}_1 - \bar{x}_2$	$n_1 = n_2 = \dfrac{(z_{\alpha/2})^2(\sigma_1^2 + \sigma_2^2)}{E^2}$	(3)
$p_1 - p_2$	$\hat{p}_1 - \hat{p}_2$	$n_1 = n_2 = \dfrac{(z_{\alpha/2})^2(p_1 q_1 + p_2 q_2)}{E^2}$	(4)

Table 9.2: Common sample size calculations.

9.4 SUMMARY OF COMMON CONFIDENCE INTERVALS

Table 9.3 presents a summary of common confidence intervals for one sample, Table 9.4 is for two samples. For each population parameter, the assumptions and formula for a $100(1-\alpha)\%$ confidence interval are given.

Parameter	Assumptions	$100(1-\alpha)\%$ Confidence interval	
μ	n large, σ^2 known, or normality, σ^2 known	$\bar{x} \pm z_{\alpha/2} \cdot \dfrac{\sigma}{\sqrt{n}}$	(1)
μ	normality, σ^2 unknown	$\bar{x} \pm t_{\alpha/2,n-1} \cdot \dfrac{s}{\sqrt{n}}$	(2)
σ^2	normality	$\left(\dfrac{(n-1)s^2}{\chi^2_{\alpha/2,n-1}}, \dfrac{(n-1)s^2}{\chi^2_{1-\alpha/2,n-1}}\right)$	(3)
p	binomial experiment, n large	$\hat{p} \pm z_{\alpha/2} \cdot \sqrt{\dfrac{\hat{p}(1-\hat{p})}{n}}$	(4)

Table 9.3: Summary of common confidence intervals: one sample.

Example 9.45: A software company conducted a survey on the size of a typical word processing file. For $n = 23$ randomly selected files, $\bar{x} = 4822$ kb and $s = 127$. Find a 95% confidence interval for the true mean size of word processing files.

Solution:

(S1) The underlying population, the size of word processing files, is assumed to be normal. The confidence interval for μ is based on a t distribution. Use formula (2) of Table 9.3 as follows.

Parameter	Assumptions	$100(1-\alpha)\%$ Confidence interval	
$\mu_1 - \mu_2$	normality, independence, σ_1^2, σ_2^2 known or n_1, n_2 large, independence, σ_1^2, σ_2^2 known	$(\bar{x}_1 - \bar{x}_2) \pm z_{\alpha/2} \cdot \sqrt{\dfrac{\sigma_1^2}{n_1} + \dfrac{\sigma_2^2}{n_2}}$	(1)
$\mu_1 - \mu_2$	normality, independence, $\sigma_1^2 = \sigma_2^2$ unknown	$(\bar{x}_1 - \bar{x}_2) \pm t_{\frac{\alpha}{2}, n_1+n_2-2} \cdot s_p \sqrt{\dfrac{1}{n_1} + \dfrac{1}{n_2}}$ $s_p^2 = \dfrac{(n_1-1)s_1^2 + (n_2-1)s_2^2}{n_1 + n_2 - 2}$	(2)
$\mu_1 - \mu_2$	normality, independence, $\sigma_1^2 \neq \sigma_2^2$ unknown	$(\bar{x}_1 - \bar{x}_2) \pm t_{\alpha/2,\nu} \cdot \sqrt{\dfrac{s_1^2}{n_1} + \dfrac{s_2^2}{n_2}}$ $\nu \approx \dfrac{\left(\dfrac{s_1^2}{n_1} + \dfrac{s_2^2}{n_2}\right)^2}{\dfrac{(s_1^2/n_1)^2}{n_1-1} + \dfrac{(s_2^2/n_2)^2}{n_2-1}}$	(3)
$\mu_1 - \mu_2$	normality, n pairs, dependence	$\bar{d} \pm t_{\alpha/2, n-1} \cdot \dfrac{s_d}{\sqrt{n}}$	(4)
σ_1^2/σ_2^2	normality, independence	$\left(\dfrac{s_1^2}{s_2^2} \cdot \dfrac{1}{F_{\frac{\alpha}{2}, n_1-1, n_2-1}},\ \dfrac{s_1^2}{s_2^2} \cdot \dfrac{1}{F_{1-\frac{\alpha}{2}, n_1-1, n_2-1}}\right)$	(5)
$p_1 - p_2$	binomial experiments, n_1, n_2 large, independence	$(\hat{p}_1 - \hat{p}_2) \pm z_{\alpha/2} \cdot \sqrt{\dfrac{\hat{p}_1(1-\hat{p}_1)}{n_1} + \dfrac{\hat{p}_2(1-\hat{p}_2)}{n_2}}$	(6)

Table 9.4: Summary of common confidence intervals: two samples.

(S2) $1 - \alpha = .95$; $\alpha = .05$; $\alpha/2 = .025$; $t_{\alpha/2, n-1} = t_{.025, 22} = 2.0739$.

(S3) $k = (2.0739)(127)/\sqrt{23} = 54.92$.

(S4) A 99% confidence interval for μ: $(\bar{x} - k, \bar{x} + k) = (4767, 4877)$.

9.5 OTHER TESTS

9.5.1 Confidence interval for medians

Find an approximate $100(1-\alpha)\%$ confidence interval for the median $\tilde{\mu}$ where n is large (based on the Wilcoxon one-sample statistic).

9.5. OTHER TESTS

n	$\alpha = .05$		$\alpha = .01$	
	k_1	k_2	k_1	k_2
7	1	20		
8	2	26		
9	4	32		
10	6	39	1	44
11	8	47	2	53
12	11	55	4	62
13	14	64	6	72
14	17	74	9	82
15	21	84	12	93
16	26	94	15	105
17	30	106	18	118
18	35	118	22	131
19	41	130	27	144
20	46	144	31	159

Table 9.5: Confidence interval for median (see section 9.5.1).

(1) Compute the order statistics $\{w_{(1)}, w_{(2)}, \ldots, w_{(N)}\}$ of the $N = \binom{n}{2} = \frac{n(n-1)}{2}$ averages $(x_i + x_j)/2$, for $1 \le i < j \le n$.

(2) Determine the critical value $z_{\alpha/2}$ such that $\text{Prob}\left[Z \ge z_{\alpha/2}\right] = \alpha/2$.

(3) Compute the constants $k_1 = \left\lfloor \frac{N}{2} - \frac{z_{\alpha/2} N}{\sqrt{3n}} \right\rfloor$ and $k_2 = \left\lceil \frac{N}{2} + \frac{z_{\alpha/2} N}{\sqrt{3n}} \right\rceil$.

(4) A $100(1-\alpha)\%$ confidence interval for $\tilde{\mu}$ is given by $(w_{(k_1)}, w_{(k_2)})$. (See Table 9.5.)

9.5.2 Difference in medians

The following technique, based on the Mann–Whitney–Wilcoxon procedure, may be used to find an approximate $100(1-\alpha)\%$ confidence interval for the difference in medians, $\tilde{\mu}_1 - \tilde{\mu}_2$. Assume the sample sizes are large and the samples are independent.

(1) Compute the order statistics $\{w_{(1)}, w_{(2)}, \ldots, w_{(N)}\}$ for the $N = n_1 n_2$ differences $x_i - y_j$, for $1 \le i \le n_1$ and $1 \le j \le n_2$.

(2) Determine the critical value $z_{\alpha/2}$ such that $\text{Prob}\left[Z \ge z_{\alpha/2}\right] = \alpha/2$.

(3) Compute the constants

$$k_1 = \left\lfloor \frac{n_1 n_2}{2} + 0.5 - z_{\alpha/2} \sqrt{\frac{n_1 n_2 (n_1 + n_2 + 1)}{12}} \right\rfloor \text{ and}$$

$$k_2 = \left\lceil \frac{n_1 n_2}{2} + 0.5 + z_{\alpha/2}\sqrt{\frac{n_1 n_2 (n_1 + n_2 + 1)}{12}} \right\rceil.$$

(4) An approximate $100(1-\alpha)\%$ confidence interval for $\tilde{\mu}_1 - \tilde{\mu}_2$ is given by $(w_{(k_1)}, w_{(k_2)})$.

9.6 FINITE POPULATION CORRECTION FACTOR

Suppose a sample of size n is taken without replacement from a (finite) population of size N. If n is large or a significant portion of the population then, intuitively, a point estimate based on this sample should be more accurate than if the population were infinite. In such cases, therefore, the standard deviation of the sample mean and the standard deviation of the sample proportion are corrected (multiplied) by the **finite population correction factor**:

$$\sqrt{\frac{N-n}{N-1}}. \qquad (9.3)$$

When constructing a confidence interval, the *critical distance* is multiplied by this function of n and N to yield a more accurate interval estimate. If the sample size is *less* than 5% of the total population, the finite population correction factor is usually not applied.

Confidence intervals constructed using the finite population correction factor:

(1) Suppose a random sample of size n is taken from a population of size N. If the population is assumed normal, the endpoints for a $100(1-\alpha)\%$ confidence interval for the population mean μ are

$$\overline{x} \pm z_{\alpha/2} \cdot \frac{s}{\sqrt{n}} \cdot \sqrt{\frac{N-n}{N-1}}. \qquad (9.4)$$

(2) In a binomial experiment, suppose a random sample of size n is taken from a population of size N. The endpoints for a $100(1-\alpha)\%$ confidence interval for the population proportion p are

$$\widehat{p} \pm z_{\alpha/2} \cdot \sqrt{\frac{\widehat{p}(1-\widehat{p})}{n}} \cdot \sqrt{\frac{N-n}{N-1}}. \qquad (9.5)$$

CHAPTER 10
Hypothesis Testing

10.1 INTRODUCTION

A hypothesis test is a formal procedure used to investigate a claim about one or more population parameters. Using the information in the sample the claim is either rejected or not rejected.

There are four parts to every hypothesis test:

(1) The **null hypothesis**, H_0, is a claim about the value of one or more population parameters; assumed to be true.

(2) The **alternative**, or (**research**), **hypothesis**, H_a, is an opposing statement; believed to be true if the null hypothesis is false.

(3) The **test statistic**, TS, is a quantity computed from the sample and used to decide whether or not to reject the null hypothesis.

(4) The **rejection region**, RR, is a set or interval of numbers selected in such a way that if the value of the test statistic lies in the rejection region the null hypothesis is rejected. One or more **critical values** separate the rejection region from the remaining values of the test statistic.

There are two error probabilities associated with hypothesis testing; they are illustrated in Table 10.1 and described below.

(1) A **type I error** occurs if the null hypothesis is rejected when it is really true. The probability of a type I error is usually denoted by α, so that Prob[type I error] $= \alpha$. Common values of α include 0.05, 0.01, and 0.001.

(2) A **type II error** occurs if the null hypothesis is accepted when it is really false. The probability of a type II error depends upon the true value of the population parameter(s) and is usually denoted by β (or $\beta(\theta)$), so that Prob[type II error] $= \beta$. The **power** of the hypothesis test is $1 - \alpha$.

Note:

(1) α is the **significance level** of the hypothesis test. The test statistic is **significant** if it lies in the rejection region.

(2) The values α and β are inversely related, that is, when α increases then β decreases, and conversely.

	Decision	
	Do not reject H_0	Reject H_0
Nature — H_0 True	Correct decision	Type I error: α
Nature — H_0 False	Type II error: β	Correct decision

Table 10.1: Hypothesis test errors.

(3) To decrease both α and β, increase the sample size.

The **p-value** is the smallest value of α (the smallest significance level) that would result in rejecting the null hypothesis. A p-value for a hypothesis test is often reported rather than whether or not the value of the test statistic lies in the rejection region.

10.1.1 Tables

Tables 10.2 and 10.3 contain hypothesis tests for one and two samples. The small numbers on the right-hand side of each table are for referencing these tests.

Null hypothesis, assumptions	Alternative hypotheses	Test statistic	Rejection regions	
$\mu = \mu_0$, n large, σ^2 known, or normality, σ^2 known	$\mu > \mu_0$	$Z = \dfrac{\overline{X} - \mu_0}{\sigma/\sqrt{n}}$	$Z \geq z_\alpha$	(1)
	$\mu < \mu_0$		$Z \leq -z_\alpha$	(2)
	$\mu \neq \mu_0$		$|Z| \geq z_{\alpha/2}$	(3)
$\mu = \mu_0$, normality, σ^2 unknown	$\mu > \mu_0$	$T = \dfrac{\overline{X} - \mu_0}{S/\sqrt{n}}$	$T \geq t_{\alpha, n-1}$	(4)
	$\mu < \mu_0$		$T \leq -t_{\alpha, n-1}$	(5)
	$\mu \neq \mu_0$		$|T| \geq t_{\alpha/2, n-1}$	(6)
$\sigma^2 = \sigma_0^2$, normality	$\sigma^2 > \sigma_0^2$	$\chi^2 = \dfrac{(n-1)S^2}{\sigma_0^2}$	$\chi^2 \geq \chi^2_{\alpha, n-1}$	(7)
	$\sigma^2 < \sigma_0^2$		$\chi^2 \leq \chi^2_{1-\alpha, n-1}$	(8)
	$\sigma^2 \neq \sigma_0^2$		$\chi^2 \leq \chi^2_{1-\alpha/2, n-1}$, or $\chi^2 \geq \chi^2_{\alpha/2, n-1}$	(9)
$p = p_0$, binomial experiment, n large	$p > p_0$	$Z = \dfrac{\widehat{p} - p_0}{\sqrt{p_0(1-p_0)/n}}$	$Z \geq z_\alpha$	(10)
	$p < p_0$		$Z \leq -z_\alpha$	(11)
	$p \neq p_0$		$|Z| \geq z_{\alpha/2}$	(12)

Table 10.2: Hypothesis tests: one sample.

Example 10.46: A breakfast cereal manufacturer claims each box is filled with 24 ounces of cereal. To check this claim, a consumer group randomly selected 17 boxes and carefully weighed the contents. The summary statistics: $\overline{x} = 23.55$ and $s = 1.5$. Is there any evidence to suggest the cereal boxes are underfilled? Use $\alpha = .05$.

10.1. INTRODUCTION

Assumptions				
Null hypothesis	Alternative hypotheses	Test statistic	Rejection regions	
n_1, n_2 large, independence, σ_1^2, σ_2^2 known, or normality, independence, σ_1^2, σ_2^2 known				
$\mu_1 - \mu_2 = \Delta_0$	$\mu_1 - \mu_2 > \Delta_0$	$Z = \dfrac{(\overline{X}_1 - \overline{X}_2) - \Delta_0}{\sqrt{\dfrac{\sigma_1^2}{n_1} + \dfrac{\sigma_2^2}{n_2}}}$	$Z \geq z_\alpha$	(1)
	$\mu_1 - \mu_2 < \Delta_0$		$Z \leq -z_\alpha$	(2)
	$\mu_1 - \mu_2 \neq \Delta_0$		$|Z| \geq z_{\alpha/2}$	(3)
normality, independence, $\sigma_1^2 = \sigma_2^2$ unknown				
$\mu_1 - \mu_2 = \Delta_0$	$\mu_1 - \mu_2 > \Delta_0$	$T = \dfrac{(\overline{X}_1 - \overline{X}_2) - \Delta_0}{S_p\sqrt{\dfrac{1}{n_1} + \dfrac{1}{n_2}}}$ $S_p = \dfrac{(n_1-1)S_1^2 + (n_2-1)S_2^2}{n_1+n_2-2}$	$T \geq t_{\alpha, n_1+n_2-2}$	(4)
	$\mu_1 - \mu_2 < \Delta_0$		$T \leq -t_{\alpha, n_1+n_2-2}$	(5)
	$\mu_1 - \mu_2 \neq \Delta_0$		$|T| \geq t_{\frac{\alpha}{2}, n_1+n_2-2}$	(6)
normality, independence, σ_1^2, σ_2^2 unknown, $\sigma_1^2 \neq \sigma_2^2$				
$\mu_1 - \mu_2 = \Delta_0$	$\mu_1 - \mu_2 > \Delta_0$	$T' = \dfrac{(\overline{X}_1 - \overline{X}_2) - \Delta_0}{\sqrt{\dfrac{S_1^2}{n_1} + \dfrac{S_2^2}{n_2}}}$ $\nu \approx \dfrac{\left(\dfrac{s_1^2}{n_1} + \dfrac{s_2^2}{n_2}\right)^2}{\dfrac{(s_1^2/n_1)^2}{n_1-1} + \dfrac{(s_2^2/n_2)^2}{n_2-1}}$	$T' \geq t_{\alpha,\nu}$	(7)
	$\mu_1 - \mu_2 < \Delta_0$		$T' \leq -t_{\alpha,\nu}$	(8)
	$\mu_1 - \mu_2 \neq \Delta_0$		$|T'| \geq t_{\alpha/2,\nu}$	(9)
normality, n pairs, dependence				
$\mu_D = \Delta_0$	$\mu_D > \Delta_0$	$T = \dfrac{\overline{D} - \Delta_0}{S_D/\sqrt{n}}$	$T \geq t_{\alpha, n-1}$	(10)
	$\mu_D < \Delta_0$		$T \leq -t_{\alpha, n-1}$	(11)
	$\mu_D \neq \Delta_0$		$|T| \geq t_{\alpha/2, n-1}$	(12)
normality, independence				
$\sigma_1^2 = \sigma_2^2$	$\sigma_1^2 > \sigma_2^2$	$F = S_1^2/S_2^2$	$F \geq F_{\alpha, n_1-1, n_2-1}$	(13)
	$\sigma_1^2 < \sigma_2^2$		$F \leq F_{1-\alpha, n_1-1, n_2-1}$	(14)
	$\sigma_1^2 \neq \sigma_2^2$		$F \leq F_{1-\frac{\alpha}{2}, n_1-1, n_2-1}$ or $F \geq F_{\frac{\alpha}{2}, n_1-1, n_2-1}$	(15)
binomial experiments, n_1, n_2 large, independence				
$p_1 = p_2 = 0$	$p_1 - p_2 > 0$	$Z = \dfrac{\widehat{p}_1 - \widehat{p}_2}{\sqrt{\widehat{p}\widehat{q}(1/n_1 + 1/n_2)}}$ $\widehat{p} = \dfrac{X_1 + X_2}{n_1 + n_2}, \quad \widehat{q} = 1 - \widehat{p}$	$Z \geq z_\alpha$	(16)
	$p_1 - p_2 < 0$		$Z \leq -z_\alpha$	(17)
	$p_1 - p_2 \neq 0$		$|Z| \geq z_{\alpha/2}$	(18)
binomial experiments, n_1, n_2 large, independence				
$p_1 - p_2 = \Delta_0$	$p_1 - p_2 > \Delta_0$	$Z = \dfrac{(\widehat{p}_1 - \widehat{p}_2) - \Delta_0}{\sqrt{\dfrac{\widehat{p}_1(1-\widehat{p}_1)}{n_1} + \dfrac{\widehat{p}_2(1-\widehat{p}_2)}{n_2}}}$	$Z \geq z_\alpha$	(19)
	$p_1 - p_2 < \Delta_0$		$Z \leq -z_\alpha$	(20)
	$p_1 - p_2 \neq \Delta_0$		$|Z| \geq z_{\alpha/2}$	(21)

Table 10.3: Hypothesis tests: two samples.

Solution:

(S1) This is a question about a population mean μ. The distribution of cereal box weights is assumed normal and the population variance is unknown. A one-sample t test is appropriate (Table 10.2, number (5)).

(S2) The four parts to the hypothesis test are:

H_0: $\mu = 24 = \mu_0$

H_a: $\mu < 24$

TS: $T = \dfrac{\overline{X} - \mu_0}{S/\sqrt{n}}$

RR: $T \leq -t_{\alpha, n-1} = -t_{.05, 16} = -1.7459$

(S3) $T = \dfrac{23.55 - 24}{1.5/\sqrt{17}} = -1.2369$

(S4) Conclusion: The value of the test statistic does not lie in the rejection region (equivalently, $p = .1170 > .05$). There is no evidence to suggest the population mean is less than 24 ounces.

Example 10.47: A newspaper article claimed the proportion of local residents in favor of a property tax increase to fund new educational programs is .45. A school board member selected 192 random residents and found 65 were in favor of the tax increase. Is there any evidence to suggest the proportion reported in the newspaper article is wrong? Use $\alpha = 0.1$.

Solution:

(S1) This is a question about a population proportion p. A binomial experiment is assumed and n is large. A one-sample test based on a Z statistic is appropriate (Table 10.2, number (12)).

(S2) The four parts to the hypothesis test are:

H_0: $p = .45 = p_0$

H_a: $p \neq .45$

TS: $Z = \dfrac{\widehat{p} - p_0}{\sqrt{p_0(1 - p_0)/n}}$

RR: $|Z| \geq z_{\alpha/2} = z_{.005} = 2.5758$

(S3) $\widehat{p} = \dfrac{65}{192} = .3385$; $Z = \dfrac{.3385 - .45}{\sqrt{(.3385)(.6615)/192}} = -3.1044$

(S4) Conclusion: The value of the test statistic lies in the rejection region (equivalently, $p = .0019 < .005$). There is evidence to suggest the true proportion of residents in favor of the property tax increase is different from .45.

Example 10.48: An automobile parts seller claims a new product when attached to an engine's air filter will significantly improve gas mileage. To test this claim, a consumer group randomly selected 10 cars and drivers. The miles per gallon for each automobile was recorded without the product and then using the new product. The summary statistics for the differences (before − after) were: $\overline{d} = -1.2$ and $s_D = 3.5$. Is there any evidence to suggest the new product improves gas mileage? Use $\alpha = .01$.

Solution:

(S1) This is a question about a difference in population means, μ_D. The data are assumed to be from a normal distribution and the observations are dependent. A paired t test is appropriate (Table 10.3, number (5)).

(S2) The four parts to the hypothesis test are:

H_0: $\mu_D = 0 = \Delta_0$

H_a: $\mu_D < 0$

TS: $T = \dfrac{\overline{D} - \Delta_0}{S_d/\sqrt{n}}$

RR: $T \leq -t_{\alpha, n-1} = -t_{.01, 9} = -2.8214$

(S3) $T = \dfrac{-1.2 - 0}{3.5/\sqrt{10}} = -1.0842$

(S4) Conclusion: The value of the test statistic does not lie in the rejection region (equivalently, $p = .1532 > .01$). There is no evidence to suggest the new product improves gas mileage.

10.2 THE NEYMAN–PEARSON LEMMA

Given the null hypothesis H_0: $\theta = \theta_0$ versus the alternative hypothesis H_a: $\theta = \theta_a$, let $L(\theta)$ be the likelihood function evaluated at θ. For a given α, the test that maximizes the power at θ_a has a rejection region determined by

$$\frac{L(\theta_0)}{L(\theta_a)} < k \tag{10.1}$$

This statistical test is the most powerful test of H_0 versus H_a.

10.3 LIKELIHOOD RATIO TESTS

Given the null hypothesis H_0: $\boldsymbol{\theta} \in \Omega_0$ versus the alternative hypothesis H_a: $\boldsymbol{\theta} \in \Omega_a$ with $\Omega_0 \cap \Omega_a = \phi$ and $\Omega = \Omega_0 \cup \Omega_a$. Let $L(\widehat{\Omega}_0)$ be the likelihood function with all unknown parameters replaced by their maximum likelihood estimators subject to the constraint $\boldsymbol{\theta} \in \Omega_0$, and let $L(\widehat{\Omega})$ be defined similarly so that $\boldsymbol{\theta} \in \Omega$. Define

$$\lambda = \frac{L(\widehat{\Omega}_0)}{L(\widehat{\Omega})}. \tag{10.2}$$

A likelihood ratio test of H_0 versus H_a uses λ as a test statistic and has a rejection region given by $\lambda \leq k$ (for $0 < k < 1$).

Under very general conditions and for large n, $-2 \ln \lambda$ has approximately a chi-square distribution with degrees of freedom equal to the number of parameters or functions of parameters assigned specific values under H_0.

10.4 GOODNESS OF FIT TEST

Let n_i be the number of observations falling into the ith category (for $i = 1, 2, \ldots, k$) and let $n = n_1 + n_2 + \cdots + n_k$.

H_0: $p_1 = p_{10}$, $p_2 = p_{20}$, ..., $p_k = p_{k0}$

H_a: $p_i \neq p_{i0}$ for at least one i

TS: $\chi^2 = \sum_{i=1}^{k} \frac{(\text{observed} - \text{estimated expected})^2}{\text{estimated expected}} = \sum_{i=1}^{k} \frac{(n_i - np_{i0})^2}{np_{i0}}$

Under the null hypothesis χ^2 has approximately a chi-square distribution with $k-1$ degrees of freedom. The approximation is satisfactory if $np_{i0} \geq 5$ for all i.

RR: $\chi^2 \geq \chi^2_{\alpha, k-1}$

Example 10.49: The bookstore at a large university stocks four brands of graphing calculators. Recent sales figures indicated 55% of all graphing calculator sales were Texas Instruments (TI), 25% were Hewlett Packard (HP), 15% were Casio, and 5% were Sharp. This semester 200 graphing calculators were sold according to the table given below. Is there any evidence to suggest the sales proportions have changed? Use $\alpha = .05$.

Calculator Sales			
TI	HP	Casio	Sharp
120	47	21	12

Solution:

(S1) There are $k = 4$ categories (of calculators) with unequal expected frequencies. The bookstore would like to determine if sales are consistent with previous results. This problem involves a goodness of fit test based on a chi-square distribution.

(S2) The four parts to the hypothesis test are:

H_0: $p_1 = .55$, $p_2 = .25$, $p_3 = .15$, $p_4 = .05$.

H_a: $p_i \neq p_{i0}$ for at least one i

TS: $\chi^2 = \sum_{i=1}^{4} \frac{(\text{observed} - \text{estimated expected})^2}{\text{estimated expected}} = \sum_{i=1}^{4} \frac{(n_i - np_{i0})^2}{np_{i0}}$

RR: $\chi^2 \geq \chi^2_{\alpha, k-1} = \chi^2_{.05, 3} = 7.8147$

(S3) $\chi^2 = \frac{(120 - 110)^2}{110} + \frac{(50 - 47)^2}{50} + \frac{(30 - 21)^2}{30} + \frac{(10 - 12)^2}{10} = 4.1891$

(S4) Conclusion: The value of the test statistic does not lie in the rejection region (equivalently, $p = .2418 > .05$). There is no evidence to suggest the proportions of graphing calculator sales have changed.

If $k = 2$, this test is equivalent to a one proportion Z test, Table 10.2, number (3). This result follows from section 6.8.3 (page 99): If Z is a standard normal random variable, then Z^2 has a chi-square distribution with 1 degree of freedom.

10.5 CONTINGENCY TABLES

The general $I \times J$ contingency table has the form:

	Treatment 1	Treatment 1	...	Treatment J	Totals
Sample 1	n_{11}	n_{12}	...	n_{1J}	$n_{1.}$
Sample 2	n_{21}	n_{22}	...	n_{2J}	$n_{2.}$
$\vdots$	$\vdots$	$\vdots$	$\ddots$	$\vdots$	$\vdots$
Sample I	n_{I1}	n_{I2}	...	n_{IJ}	$n_{I.}$
Totals	$n_{.1}$	$n_{.2}$	...	$n_{.J}$	n

where $n_{k.} = \sum_{j=1}^{J} n_{kj}$ and $n_{.k} = \sum_{i=1}^{I} n_{ik}$. If complete independence is assumed, then the probability of any specific configuration, given the row and column totals $\{n_{.k}, n_{k.}\}$, is

$$\text{Prob}[n_{11}, \ldots, n_{IJ} \mid n_{1.}, \ldots, n_{.J}] = \frac{(\Pi_i^I n_{i.}!)(\Pi_j^J n_{.j}!)}{n! \, \Pi_i^I \Pi_j^J n_{ij}!} \quad (10.3)$$

Let a contingency table contain I rows and J columns, let n_{ij} be the count in the $(i,j)^{\text{th}}$ cell, and let $\hat{\epsilon}_{ij}$ be the estimated expected count in that cell. The test statistic is

$$\chi^2 = \sum_{\text{all cells}} \frac{(\text{observed} - \text{estimated expected})^2}{\text{estimated expected}} = \sum_{i=1}^{I} \sum_{j=1}^{J} \frac{(n_{ij} - \hat{\epsilon}_{ij})^2}{\hat{\epsilon}_{ij}} \quad (10.4)$$

where

$$\hat{\epsilon}_{ij} = \frac{(i^{\text{th}} \text{ row total})(j^{\text{th}} \text{ column total})}{\text{grand total}} = \frac{n_{i.} n_{.j}}{n} \quad (10.5)$$

Under the null hypothesis χ^2 has approximately a chi-square distribution with $(I-1)(J-1)$ degrees of freedom. The approximation is satisfactory if $\hat{\epsilon}_{ij} \geq 5$ for all i and j.

Example 10.50: Recent reports indicate meals served during flights are rated similar regardless of airline. A survey given to randomly selected passengers asked each to rate the quality of in-flight meals. The results are given in the table below.

	Airline			
	A	B	C	D
Poor	42	35	22	23
Acceptable	50	75	33	28
Good	10	17	21	18

Is there any evidence to suggest the quality of meals differs by airline? Use $\alpha = .01$.

Solution:

(S1) The contingency table has $I = 3$ rows and $J = 4$ columns. To determine if the meal ratings differ by airline, a contingency table analysis is appropriate. The test statistic is based on a chi-square distribution.

(S2) The four parts to the hypothesis test are:

H_0: Airline and meal ratings are independent

H_a: Airline and meal ratings are dependent

TS: $\chi^2 = \sum_{i=1}^{3}\sum_{j=1}^{4} \frac{(n_{ij} - \hat{e}_{ij})^2}{\hat{e}_{ij}}$

RR: $\chi^2 \geq \chi^2_{.01,6} = 18.5476$

(S3) $\chi^2 = \frac{(42 - 33.27)^2}{33.27} + \frac{(35 - 41.43)^2}{41.43} + \frac{(22 - 24.79)^2}{24.79} + \frac{(23 - 22.51)^2}{22.51}$
$+ \frac{(50 - 50.73)^2}{50.73} + \frac{(75 - 63.16)^2}{63.16} + \frac{(33 - 37.80)^2}{37.80} + \frac{(28 - 34.32)^2}{34.32}$
$+ \frac{(10 - 18.00)^2}{18.00} + \frac{(17 - 22.41)^2}{22.41} + \frac{(21 - 13.41)^2}{13.41} + \frac{(18 - 12.18)^2}{12.18}$
$= 19.553$

(S4) The value of the test statistic lies in the rejection region (i.e., $p = .003 < .01$). There is evidence to suggest the meal rating proportions differ by airline.

10.6 TEST OF SIGNIFICANCE IN 2 × 2 CONTINGENCY TABLES

A 2×2 contingency table (see section 10.5) is a special case that occurs often. Suppose n elements are simultaneously classified as having either property 1 or 2 and as having property I or II. The 2×2 contingency table may be written as:

	I	II	Totals
1	a	$A-a$	A
2	b	$B-b$	B
Totals	r	$n-r$	n

If the marginal totals (r, A, and B) are fixed, the probability of a given configuration may be written as

$$f(a \mid r, A, B) = \frac{\binom{A}{a}\binom{B}{b}}{\binom{n}{r}} = \frac{A!\,B!\,r!\,(n-r)!}{n!\,a!\,b!\,(A-a)!\,(B-b)!} \quad (10.6)$$

Given a, A, B, and equation (10.6) a *critical value* of r may be determined so that $f(a \mid r, A, B)$ is a desired probability. The number of elements with properties 1 and I is then $b = r - a$.

Example 10.51: In order to compare the probability of a success in two populations, the following 2×2 contingency table was obtained.

	Success	Failure	Totals
Sample from population 1	7	2	9
Sample from population 2	3	3	6
Totals	10	5	15

10.7. CRITICAL VALUES FOR TESTING OUTLIERS

Is there any evidence to suggest the two population proportions are different? Use $\alpha = .05$.

Solution:

(S1) For the given values of a, A, B, and unknown r, the contingency table is given by

	Success	Failure	Totals
Sample from population 1	7	2	9
Sample from population 2	b	$6-b$	6
Totals	r	$15-r$	15

(S2) For $7 \leq r \leq 13$ the conditional probability $f(a \mid r, A, B)$ is given by

r	7	8	9	10	11	12	13
$f(7 \mid r, 9, 6)$	0.0056	0.034	0.11	0.24	0.40	0.47	0.34

(S3) Using this table, the largest value of r for which the probability of observing the given value of a is less than $\alpha = 0.05$ is $r = 8$. The actual probability of observing $a = 7$ is 0.034.

(S4) Therefore, the critical value is $b = r - a = 8 - 7 = 1$. If, in the actual data, $b \leq 1$ then the null hypothesis H_0: $p_1 = p_2$ would be rejected. In this example (from the given data) the value $b = 3$ was observed.

(S5) Conclusion: The value of the test statistic does not lie in the rejection region. There is no evidence to suggest the population proportions are different.

10.7 CRITICAL VALUES FOR TESTING OUTLIERS

Tests for outliers may be based on the largest deviation $\max_{i=1,2,\ldots}(x_i - \bar{x})$ of the observations from their mean (which has to be normalized by the standard deviation or an estimate of the standard deviation). An alternative technique is to look at ratios of approximations to the range.

(a) To determine if the smallest element in a sample, $x_{(1)}$, is an outlier compute

$$r_{10} = \frac{x_{(2)} - x_{(1)}}{x_{(n)} - x_{(1)}} \qquad (10.7)$$

Equivalently, to determine if the largest element in a sample, $x_{(n)}$, is an outlier compute

$$r_{10} = \frac{x_{(n)} - x_{(n-1)}}{x_{(n)} - x_{(1)}} \qquad (10.8)$$

(b) To determine if the smallest element in a sample, $x_{(1)}$, is an outlier, and the value $x_{(n)}$ is not to be used, then compute

$$r_{11} = \frac{x_{(2)} - x_{(1)}}{x_{(n-1)} - x_{(1)}} \qquad (10.9)$$

Equivalently, to determine if the largest element in a sample, $x_{(n)}$, is an outlier, without using the value $x_{(1)}$, compute

$$r_{11} = \frac{x_{(n)} - x_{(n-1)}}{x_{(n)} - x_{(2)}} \qquad (10.10)$$

The following tables contain critical values for r_{10}, r_{11}, and r_{20}. See W. J. Dixon, *Annals of Mathematical Statistics*, **22**, 1951, pages 68–78.

Percentage values for r_{10}   (Prob$[r_{10} > R] = \alpha$)

n	$\alpha = .005$	.01	.02	.05	.10	.50	.90	.95
3	.994	.988	.976	.941	.886	.500	.114	.059
4	.926	.889	.846	.745	.679	.324	.065	.033
5	.821	.780	.729	.642	.557	.250	.048	.023
6	.740	.698	.644	.560	.482	.210	.038	.018
7	.680	.637	.586	.507	.434	.184	.032	.016
8	.634	.590	.543	.468	.399	.166	.029	.014
9	.598	.555	.510	.437	.370	.152	.026	.013
10	.568	.527	.483	.412	.349	.142	.025	.012
15	.475	.438	.399	.338	.285	.111	.019	.010
20	.425	.391	.356	.300	.252	.096	.017	.008
25	.393	.362	.329	.277	.230	.088	.015	.008
30	.372	.341	.309	.260	.215	.082	.014	.007

Percentage values for r_{11}   (Prob$[r_{11} > R] = \alpha$)

n	$\alpha = .005$	.01	.02	.05	.10	.50	.90	.95
4	.995	.991	.981	.955	.910	.554	.131	.069
5	.937	.916	.876	.807	.728	.369	.078	.039
6	.839	.805	.763	.689	.609	.288	.056	.028
7	.782	.740	.689	.610	.530	.241	.045	.022
8	.725	.683	.631	.554	.479	.210	.037	.019
9	.677	.635	.587	.512	.441	.189	.033	.016
10	.639	.597	.551	.477	.409	.173	.030	.014
15	.522	.486	.445	.381	.323	.129	.023	.011
20	.464	.430	.392	.334	.282	.110	.019	.010
25	.426	.394	.359	.394	.255	.098	.017	.009
30	.399	.369	.336	.283	.236	.090	.016	.008

CHAPTER 11

Regression Analysis

11.1 SIMPLE LINEAR REGRESSION

Let $(x_1, y_1), (x_2, y_2), \ldots, (x_n, y_n)$ be n pairs of observations such that y_i is an observed value of the random variable Y_i. Assume there exist constants β_0 and β_1 such that

$$Y_i = \beta_0 + \beta_1 x_i + \epsilon_i \qquad (11.1)$$

where $\epsilon_1, \epsilon_2, \ldots, \epsilon_n$ are independent, normal random variables having mean 0 and variance σ^2.

Assumptions	
In terms of ϵ_i's	In terms of Y_i's
ϵ_i's are normally distributed	Y_i's are normally distributed
$E[\epsilon_i] = 0$	$E[Y_i] = \beta_0 + \beta_1 x_i$
$\text{Var}[\epsilon_i] = \sigma^2$	$\text{Var}[Y_i] = \sigma^2$
$\text{Cov}[\epsilon_i, \epsilon_j] = 0, \ i \neq j$	$\text{Cov}[Y_i, Y_j] = 0, \ i \neq j$

Principle of least squares: The sum of squared deviations about the true regression line is

$$S(\beta_0, \beta_1) = \sum_{i=1}^{n} [y_i - (\beta_0 + \beta_1 x_i)]^2. \qquad (11.2)$$

The point estimates of β_0 and β_1, denoted by $\widehat{\beta}_0$ and $\widehat{\beta}_1$, are those values that minimize $S(\beta_0, \beta_1)$. The estimates $\widehat{\beta}_0$ and $\widehat{\beta}_1$ are called the **least squares estimates**. The estimated regression line or least squares line is $\widehat{y} = \widehat{\beta}_0 + \widehat{\beta}_1 x$.

The **normal equations** for $\widehat{\beta}_0$ and $\widehat{\beta}_1$ are

$$\begin{aligned} \left(\sum_{i=1}^{n} y_i\right) &= \widehat{\beta}_0 \, n + \widehat{\beta}_1 \left(\sum_{i=1}^{n} x_i\right) \\ \left(\sum_{i=1}^{n} x_i y_i\right) &= \widehat{\beta}_0 \left(\sum_{i=1}^{n} x_i\right) + \widehat{\beta}_1 \left(\sum_{i=1}^{n} x_i^2\right) \end{aligned} \qquad (11.3)$$

Notation:

$$S_{xx} = \sum_{i=1}^{n}(x_i - \overline{x})^2 = \sum_{i=1}^{n} x_i^2 - \frac{1}{n}\left(\sum_{i=1}^{n} x_i\right)^2$$

$$S_{yy} = \sum_{i=1}^{n}(y_i - \overline{y})^2 = \sum_{i=1}^{n} y_i^2 - \frac{1}{n}\left(\sum_{i=1}^{n} y_i\right)^2$$

$$S_{xy} = \sum_{i=1}^{n}(x_i - \overline{x})(y_i - \overline{y})^2 = \sum_{i=1}^{n} x_i y_i - \frac{1}{n}\left(\sum_{i=1}^{n} x_i\right)\left(\sum_{i=1}^{n} y_i\right)$$

11.1.1 Least squares estimates

$$\widehat{\beta}_1 = \frac{S_{xy}}{S_{xx}} = \frac{n\left(\sum_{i=1}^{n} x_i y_i\right) - \left(\sum_{i=1}^{n} x_i\right)\left(\sum_{i=1}^{n} y_i\right)}{n\left(\sum_{i=1}^{n} x_i^2\right) - \left(\sum_{i=1}^{n} x_i\right)^2} \qquad (11.4)$$

$$\widehat{\beta}_0 = \frac{\sum_{i=1}^{n} y_i - \widehat{\beta}_1 \sum_{i=1}^{n} x_i}{n} = \overline{y} - \widehat{\beta}_1 \overline{x}$$

The ith **predicted (fitted) value:** $\widehat{y}_i = \widehat{\beta}_0 + \widehat{\beta}_1 x_i$ (for $i = 1, 2, \ldots, n$).
The ith **residual:** $e_i = y_i - \widehat{y}_i$ (for $i = 1, 2, \ldots, n$).
Properties:

(1) $\mathrm{E}[\widehat{\beta}_1] = \beta_1, \quad \mathrm{Var}[\widehat{\beta}_1] = \dfrac{\sigma^2}{\sum\limits_{i=1}^{n}(x_i - \overline{x})^2} = \dfrac{\sigma^2}{S_{xx}}$

(2) $\mathrm{E}[\widehat{\beta}_0] = \beta_0, \quad \mathrm{Var}[\widehat{\beta}_0] = \dfrac{\sigma^2 \sum\limits_{i=1}^{n} x_i}{n \sum\limits_{i=1}^{n}(x_i - \overline{x})^2} = \dfrac{\sigma^2 \sum\limits_{i=1}^{n} x_i}{n S_{xx}} = \dfrac{\sigma^2 \overline{x}}{S_{xx}}$

(3) $\widehat{\beta}_0$ and $\widehat{\beta}_1$ are normally distributed.

11.1. SIMPLE LINEAR REGRESSION

11.1.2 Sum of squares

$$\underbrace{\sum_{i=1}^{n}(y_i - \bar{y})^2}_{\text{SST}} = \underbrace{\sum_{i=1}^{n}(\hat{y}_i - \bar{y})^2}_{\text{SSR}} + \underbrace{\sum_{i=1}^{n}(y_i - \hat{y}_i)^2}_{\text{SSE}}$$

SST = total sum of squares = S_{yy}

SSR = sum of squares due to regression = $\hat{\beta}_2 S_{xy}$

SSE = sum of squares due to error

$$= \sum_{i=1}^{n}[y_i - (\hat{\beta}_0 + \hat{\beta}_1 x_i)]^2 = \sum_{i=1}^{n} y_i^2 - \hat{\beta}_0 \sum_{i=1}^{n} y_i - \hat{\beta}_1 \sum_{i=1}^{n} x_i y_i$$

$$= S_{yy} - 2\hat{\beta}_1 S_{xy} + \hat{\beta}_1^2 S_{xx} = S_{yy} - \hat{\beta}_1^2 S_{xx} = S_{yy} - \hat{\beta}_1 S_{xy}$$

$$\hat{\sigma}^2 = s^2 = \frac{\text{SSE}}{n-2}, \quad E\left[S^2\right] = \sigma^2$$

Sample coefficient of determination: $r^2 = \dfrac{\text{SSR}}{\text{SST}} = 1 - \dfrac{\text{SSE}}{\text{SST}}$

11.1.3 Inferences concerning the regression coefficients

The parameter $\hat{\beta}_1$

(1) $T = \dfrac{\hat{\beta}_1 - \beta_1}{S/\sqrt{S_{xx}}} = \dfrac{\hat{\beta}_1 - \beta_1}{S_{\hat{\beta}_1}}$

has a t distribution with $n - 2$ degrees of freedom, where $S_{\hat{\beta}_1} = S/\sqrt{S_{xx}}$ is an estimate for the standard deviation of $\hat{\beta}_1$.

(2) A $100(1-\alpha)\%$ confidence interval for β_1 has as endpoints
$$\hat{\beta}_1 \pm t_{\alpha/2, n-2} \cdot s_{\hat{\beta}_1}$$

(3) Hypothesis test:

Null hypothesis	Alternative hypotheses	Test statistic	Rejection regions	
$\beta_1 = \beta_{10}$	$\beta_1 > \beta_{10}$	$T = \dfrac{\hat{\beta}_1 - \beta_{10}}{S_{\hat{\beta}_1}}$	$T \geq t_{\alpha, n-2}$	(1)
	$\beta_1 < \beta_{10}$		$T \leq -t_{\alpha, n-2}$	(2)
	$\beta_1 \neq \beta_{10}$		$\lvert T \rvert \geq t_{\alpha/2, n-2}$	(3)

The parameter $\widehat{\beta}_0$

(1) $T = \dfrac{\widehat{\beta}_0 - \beta_0}{S\sqrt{\sum\limits_{i=1}^{n} x_i^2 / nS_{xx}}} = \dfrac{\widehat{\beta}_0 - \beta_0}{S_{\widehat{\beta}_0}}$

has a t distribution with $n - 2$ degrees of freedom, where $S_{\widehat{\beta}_0}$ denotes the estimate for the standard deviation of $\widehat{\beta}_0$.

(2) A $100(1 - \alpha)\%$ confidence interval for β_1 has as endpoints

$$\widehat{\beta}_1 \pm t_{\alpha/2, n-2} \cdot s_{\widehat{\beta}_0}$$

(3) Hypothesis test:

Null hypothesis	Alternative hypotheses	Test statistic	Rejection regions	
$\beta_0 = \beta_{00}$	$\beta_0 > \beta_{00}$	$T = \dfrac{\widehat{\beta}_0 - \beta_{00}}{S_{\widehat{\beta}_0}}$	$T \geq t_{\alpha, n-2}$	(1)
	$\beta_0 < \beta_{00}$		$T \leq -t_{\alpha, n-2}$	(2)
	$\beta_0 \neq \beta_{00}$		$|T| \geq t_{\alpha/2, n-2}$	(3)

11.1.4 The mean response

The mean response of Y given $x = x_0$ is $\mu_{Y|x_0} = \beta_0 + \beta_1 x_0$. The random variable $\widehat{Y}_0 = \widehat{\beta}_0 + \widehat{\beta}_1 x_0$ is used to estimate $\mu_{Y|x_0}$.

(1) $E[\widehat{Y}_0] = \beta_0 + \beta_1 x_0$

(2) $\text{Var}[\widehat{Y}_0] = \sigma^2 \left[\dfrac{1}{n} + \dfrac{(x_0 - \bar{x})^2}{S_{xx}} \right]$

(3) $\widehat{Y}_0$ has a normal distribution.

(4) $T = \dfrac{\widehat{Y}_0 - \mu_{Y|x_0}}{S\sqrt{(1/n) + [(x_0 - \bar{x})^2/S_{xx}]}} = \dfrac{\widehat{Y}_0 - \mu_{Y|x_0}}{S_{\widehat{Y}_0}}$

has a t distribution with $n - 2$ degrees of freedom, where $S_{\widehat{Y}_0}$ denotes the estimate for the standard deviation of $\widehat{Y}_0$.

(5) A $100(1 - \alpha)\%$ confidence interval for $\mu_{Y|x_0}$ has as endpoints

$$\widehat{y}_0 \pm t_{\alpha/2, n-2} \cdot s_{\widehat{Y}_0}.$$

(6) Hypothesis test:

Null hypothesis	Alternative hypotheses	Test statistic	Rejection regions	
$\beta_0 + \beta_1 x_0 = y_0 = \mu_0$	$y_0 > \mu_0$	$T = \dfrac{\widehat{Y}_0 - \mu_0}{S_{\widehat{Y}_0}}$	$T \geq t_{\alpha, n-2}$	(1)
	$y_0 < \mu_0$		$T \leq -t_{\alpha, n-2}$	(2)
	$y_0 \neq \mu_0$		$|T| \geq t_{\alpha/2, n-2}$	(3)

11.1.5 Prediction interval

A prediction interval for a value y_0 of the random variable $Y_0 = \beta_0 + \beta_1 x_0 + \epsilon_0$ is obtained by considering the random variable $\widehat{Y}_0 - Y_0$.

(1) $E[\widehat{Y}_0 - Y_0] = 0$

(2) $\text{Var}[\widehat{Y}_0 - Y_0] = \sigma^2 \left[1 + \dfrac{1}{n} + \dfrac{(x_0 - \bar{x})^2}{S_{xx}} \right]$

(3) $\widehat{Y}_0 - Y_0$ has a normal distribution.

(4) $T = \dfrac{\widehat{Y}_0 - Y_0}{S\sqrt{1 + (1/n) + [(x_0 - \bar{x})^2/S_{xx}]}} = \dfrac{\widehat{Y}_0 - Y_0}{S_{\widehat{Y}_0 - Y_0}}$

has a t distribution with $n - 2$ degrees of freedom.

(5) A $100(1 - \alpha)\%$ prediction interval for y_0 has as endpoints
$$\widehat{y}_0 \pm t_{\alpha/2, n-2} \cdot s_{\widehat{Y}_0 - Y_0}$$

11.1.6 Analysis of variance table

Source of variation	Sum of squares	Degrees of freedom	Mean square	Computed F
Regression	SSR	1	$\text{MSR} = \dfrac{\text{SSR}}{1}$	MSR/MSE
Error	SSE	$n - 2$	$\text{MSE} = \dfrac{\text{SSE}}{n - 2}$	
Total	SST	$n - 1$		

Hypothesis test of significant regression:

Null hypothesis	Alternative hypothesis	Test statistic	Rejection region
$\beta_1 = 0$	$\beta_1 \neq 0$	$F = \text{MSR/MSE}$	$F \geq F_{\alpha, 1, n-2}$

11.1.7 Test for linearity of regression

Suppose there are k distinct values of x, $\{x_1, x_2, \ldots, x_k\}$, n_i observations for x_i, and $n = n_1 + n_2 + \cdots + n_k$.

Definitions:
(1) y_{ij} = the jth observation on the random variable Y_i.

(2) $T_i = \sum_{j=1}^{n_i} y_{ij}$, $\bar{y}_{i.} = T_i/n_i$

(3) SSPE = sum of squares due to pure error
$$= \sum_{i=1}^{k}\sum_{j=1}^{n_i}(y_{ij} - \bar{y}_{i.})^2 = \sum_{i=1}^{k}\sum_{j=1}^{n_i} y_{ij}^2 - \sum_{i=1}^{k}\frac{T_i^2}{n_i}$$

(4) SSLF = Sum of squares due to lack of fit = SSE − SSPE

Hypothesis test:

Null hypothesis	Alternative hypothesis	Test statistic	Rejection region
Linear regression	Lack of fit	$F = \dfrac{\text{SSLF}/(k-2)}{\text{SSPE}/(n-k)}$	$F \geq F_{\alpha, k-2, n-k}$

11.1.8 Sample correlation coefficient

The sample correlation coefficient is a measure of linear association and is defined by

$$r = \hat{\beta}_1 \sqrt{\frac{S_{xx}}{S_{yy}}} = \frac{S_{xy}}{\sqrt{S_{xx}S_{yy}}}. \tag{11.5}$$

Hypothesis tests:

Null hypothesis	Alternative hypothesis	Test statistic	Rejection region	
$\rho = 0$	$\rho > 0$		$T \geq t_{\alpha, n-2}$	(1)
	$\rho < 0$	$T = \dfrac{R\sqrt{n-2}}{\sqrt{1-R^2}} = \dfrac{\hat{\beta}_1}{S_{\hat{\beta}_1}}$	$T \leq -t_{\alpha, n-2}$	(2)
	$\rho \neq 0$		$\|T\| \geq t_{\alpha/2, n-2}$	(3)

If X and Y have a bivariate normal distribution:				
$\rho = \rho_0$	$\rho > \rho_0$		$Z \geq z_\alpha$	(4)
	$\rho < \rho_0$	$Z = \dfrac{\sqrt{n-3}}{2}$ $\times \ln\left[\dfrac{(1+R)(1-\rho_0)}{(1-R)(1+\rho_0)}\right]$	$Z \leq -z_\alpha$	(5)
	$\rho \neq \rho_0$		$\|Z\| \geq z_{\alpha/2}$	(6)

11.1.9 Example

Example 11.52: A recent study at a manufacturing facility examined the relationship between the noon temperature (Fahrenheit) inside the plant and the number of defective

11.1. SIMPLE LINEAR REGRESSION

items produced during the day shift. The data are given in the following table.

Noon temperature (x)	Number defective (y)	Noon temperature (x)	Number defective (y)
68	27	74	48
78	52	65	33
71	39	72	45
69	22	73	51
66	21	67	29
75	66	77	65

(1) Find the regression equation using temperature as the independent variable and construct the anova table.
(2) Test for a significant regression. Use $\alpha = .05$.
(3) Find a 95% confidence interval for the mean number of defective items produced when the temperature is 68°F.
(4) Find a 99% prediction interval for a temperature of 75°F.

Solution:

(S1) $\widehat{\beta}_1 = \dfrac{S_{xy}}{S_{xx}} = \dfrac{640.5}{204.25} = 3.1359$

$\widehat{\beta}_0 = \bar{y} - \widehat{\beta}_1 \bar{x} = 41.5 - (3.1359)(71.25) = -181.93$

Regression line: $\widehat{y} = -181.93 + 3.1359x$

(S2)

Source of variation	Sum of squares	Degrees of freedom	Mean square	Computed F
Regression	2008.5	1	2008.5	31.16
Error	644.5	10	64.4	
Total	2653.0	11		

(S3) Hypothesis test of significant regression:

$H_0: \beta_1 = 0$

$H_a: \beta_1 \neq 0$

TS: $F = \text{MSR}/\text{MSE}$

RR: $F \geq F_{.05,1,10} = 4.96$

Conclusion: The value of the test statistic ($F = 31.16$) lies in the rejection region. There is evidence to suggest a significant regression. Note: this test is equivalent to the t test in section 11.1.3 with $\beta_{10} = 0$.

(S4) A 95% confidence interval for $\mu_{Y|68}$:

$\widehat{y}_0 \pm t_{.025,10} \cdot s_{\widehat{Y}_0} = 31.31 \pm (2.228)(2.9502) = (24.74, 37.88)$

(S5) A 99% prediction interval for $y_0 = -181.93 + (3.1359)(75) = 53.26$

$\widehat{y}_0 \pm t_{.005,10} \cdot s_{\widehat{Y}_0 - Y_0} = 53.26 \pm (3.169)(8.6172) = (25.95, 80.56)$

11.2 MULTIPLE LINEAR REGRESSION

Let there be n observations of the form $(x_{1i}, x_{2i}, \ldots, x_{ki}, y_i)$ such that y_i is an observed value of the random variable Y_i. Assume there exist constants $\beta_0, \beta_1, \ldots, \beta_k$ such that

$$Y_i = \beta_0 + \beta_1 x_{1i} + \cdots + \beta_k x_{ki} + \epsilon_i \qquad (11.6)$$

where $\epsilon_1, \epsilon_2, \ldots, \epsilon_n$ are independent, normal random variables having mean 0 and variance σ^2.

Notation:

Let $\mathbf{Y}$ be the random vector of responses, $\mathbf{y}$ be the vector of observed responses, $\boldsymbol{\beta}$ be the vector of regression coefficients, $\boldsymbol{\epsilon}$ be the vector of random errors, and let $\mathbf{X}$ be the design matrix:

$$\mathbf{Y} = \begin{bmatrix} Y_1 \\ Y_2 \\ \vdots \\ Y_n \end{bmatrix} \quad \mathbf{y} = \begin{bmatrix} y_1 \\ y_2 \\ \vdots \\ y_n \end{bmatrix} \quad \boldsymbol{\beta} = \begin{bmatrix} \beta_0 \\ \beta_1 \\ \vdots \\ \beta_k \end{bmatrix} \quad \boldsymbol{\epsilon} = \begin{bmatrix} \epsilon_1 \\ \epsilon_2 \\ \vdots \\ \epsilon_n \end{bmatrix} \quad \mathbf{X} = \begin{bmatrix} 1 & x_{11} & x_{21} & \cdots & x_{k1} \\ 1 & x_{12} & x_{22} & \cdots & x_{k2} \\ \vdots & \vdots & \vdots & \ddots & \vdots \\ 1 & x_{1n} & x_{2n} & \cdots & x_{kn} \end{bmatrix} \qquad (11.7)$$

The model can now be written as $\mathbf{Y} = \mathbf{X}\boldsymbol{\beta} + \boldsymbol{\epsilon}$ where $\boldsymbol{\epsilon} \sim N_n(\mathbf{0}, \sigma^2 \mathbf{I}_n)$ or equivalently $\mathbf{Y} \sim N_n(\mathbf{X}\boldsymbol{\beta}, \sigma^2 \mathbf{I}_n)$.

Principle of least squares: The sum of squared deviations about the true regression line is

$$S(\boldsymbol{\beta}) = \sum_{i=1}^{n} [y_i - (\beta_0 + \beta_1 x_{1i} + \cdots + \beta_k x_{ki})]^2 = \|\mathbf{y} - \mathbf{X}\boldsymbol{\beta}\|^2. \qquad (11.8)$$

The vector $\widehat{\boldsymbol{\beta}} = [\widehat{\beta}_0, \widehat{\beta}_1, \ldots, \widehat{\beta}_k]^T$ that minimizes $S(\boldsymbol{\beta})$ is the vector of least squares estimates. The estimated regression line or least squares line is $y = \widehat{\beta}_0 + \widehat{\beta}_1 x_1 + \cdots + \widehat{\beta}_k x_k$.

The normal equations may be written as $(\mathbf{X}^T \mathbf{X}) \widehat{\boldsymbol{\beta}} = \mathbf{X}^T \mathbf{y}$.

11.2.1 Least squares estimates

If the matrix $\mathbf{X}^T \mathbf{X}$ is non-singular, then $\widehat{\boldsymbol{\beta}} = (\mathbf{X}^T \mathbf{X})^{-1} \mathbf{X}^T \mathbf{y}$.

The ith predicted (fitted) value: $\widehat{y}_i = \widehat{\beta}_0 + \widehat{\beta}_1 x_{1i} + \cdots + \widehat{\beta}_k x_{ki}$ (for $i = 1, 2, \ldots, n$), $\widehat{\mathbf{y}} = \mathbf{X}\widehat{\boldsymbol{\beta}}$.

The ith residual: $e_i = y_i - \widehat{y}_i$, $i = 1, 2, \ldots, n$, $\mathbf{e} = \mathbf{y} - \widehat{\mathbf{y}}$.

Properties: For $i = 0, 1, 2, \ldots, k$ and $j = 0, 1, 2, \ldots, k$:

(1) $E[\widehat{\beta}_i] = \beta_i$.

(2) $\text{Var}[\widehat{\beta}_i] = c_{ii} \sigma^2$, where c_{ij} is the value in the ith row and j^{th} column of the matrix $(\mathbf{X}^T \mathbf{X})^{-1}$.

(3) $\widehat{\beta}_i$ is normally distributed.

(4) $\text{Cov}[\widehat{\beta}_i, \widehat{\beta}_j] = c_{ij} \sigma^2$, $i \neq j$.

CHAPTER 12

Nonparametric Statistics

Nonparametric, or **distribution-free**, statistical procedures generally assume very little about the underlying population(s). The test statistic used in each procedure is usually easy to compute and may involve qualitative measurements or measurements made on an ordinal scale.

If both a parametric and nonparametric test are applicable, the nonparametric test is less efficient because it does not utilize all of the information in the sample. A larger sample size is required in order for the nonparametric test to have the same probability of a type II error.

12.1 FRIEDMAN TEST FOR RANDOMIZED BLOCK DESIGN

Assumptions: Let there be k independent random samples (treatments) from continuous distributions and b blocks.

Hypothesis test:

H_0: the k samples are from identical populations.

H_a: at least two of the populations differ in location.

Rank each observation from 1 (smallest) to k (largest) within each block. Equal observations are assigned the mean rank for their positions. Let R_i be the rank sum of the ith sample (treatment).

TS: $F_r = \left(\dfrac{12}{bk(k+1)} \sum_{i=1}^{k} R_i^2 \right) - 3b(k+1)$

RR: $F_r \geq \chi^2_{\alpha, k-1}$

12.2 KENDALL'S RANK CORRELATION COEFFICIENT

Given two sets containing ranked elements of the same size, consider each of the $\binom{n}{2} = \dfrac{n(n-1)}{2}$ pairs of elements from within each set. Associate with each pair (a) a score of $+1$ if the relative ranking of both samples is the same, or (b) a score of -1 if the relative rankings are different. Kendall's score, S_t, is defined as the total of these $\binom{n}{2}$ individual scores. S_t will have a maximum value of $\dfrac{n(n-1)}{2}$ if the two rankings are identical and a minimum value of

$-\frac{n(n-1)}{2}$ if the sets are ranked in exactly the opposite order. Kendall's Tau is defined as

$$\tau = S_t \bigg/ \left(\frac{n(n-1)}{2}\right) \tag{12.1}$$

and has the range $-1 \leq \tau \leq 1$.

The table on page 167 may be used to determine the exact probability associated with an occurrence (one-tailed) of a specific value of S_t. In this case the null hypothesis is the existence of an association between the two sets as extreme as an observed S_t. The tabled value is the probability that S_t is equaled or exceeded.

Example 12.53: Consider the sets of ranked elements: $\mathbf{a} = \{4, 12, 6, 10\}$ and $\mathbf{b} = \{8, 7, 16, 2\}$. Kendall's score is $S_t = 0$ for these sets since

For the $(1,2)$ term (with $a_1 < a_2$ and $b_1 > b_2$) score is -1
For the $(1,3)$ term (with $a_1 < a_3$ and $b_1 < b_3$) score is $+1$
For the $(1,4)$ term (with $a_1 < a_4$ and $b_1 > b_4$) score is -1
For the $(2,3)$ term (with $a_2 > a_3$ and $b_2 > b_3$) score is $+1$
For the $(2,4)$ term (with $a_2 > a_4$ and $b_2 > b_4$) score is $+1$
For the $(3,4)$ term (with $a_3 < a_4$ and $b_3 > b_4$) score is -1
Total is 0

Using the table on page 167 with $n = 4$ we find Prob $[S_t \geq 0] = .625$. That is, an S_t value of 0 or larger would be expected 62.5% of the time.

12.2.1 Tables for Kendall rank correlation coefficient

The following table may be used to determine the exact probability associated with an occurrence (one-tailed) of a specific value of S_t. In this case the null hypothesis is the existence of an association between the two sets as extreme as an observed S_t. The tabled value is the probability that S_t is equaled or exceeded.

12.3. KOLMOGOROV–SMIRNOFF TESTS

Distribution of Kendall's rank correlation coefficient in random rankings

S_t	$n=3$	4	5	6	7	8	9	10
0	0.5000	0.6250	0.5917	0.5000	0.5000	0.5476	0.5403	0.5000
2	0.1667	0.3750	0.4083	0.3597	0.3863	0.4524	0.4597	0.4309
4		0.1667	0.2417	0.2347	0.2810	0.3598	0.3807	0.3637
6		0.0417	0.1167	0.1361	0.1907	0.2742	0.3061	0.3003
8			0.0417	0.0681	0.1194	0.1994	0.2384	0.2422
10			0.0083	0.0278	0.0681	0.1375	0.1792	0.1904
12				0.0083	0.0345	0.0894	0.1298	0.1456
14				0.0014	0.0151	0.0543	0.0901	0.1082
16					0.0054	0.0305	0.0597	0.0779
18					0.0014	0.0156	0.0376	0.0542
20					0.0002	0.0071	0.0223	0.0363
22						0.0028	0.0124	0.0233
24						0.0009	0.0063	0.0143
26						0.0002	0.0029	0.0083
28							0.0012	0.0046
30							0.0004	0.0023
32							0.0001	0.0011
34								0.0005
36								0.0002
38								0.0001

Note that each distribution is symmetric about $S_t = 0$: e.g., for $n = 4$, $\text{Prob}[S_t = 2] = \text{Prob}[S_t = -2] = 0.375$. Note also that S_t can only assume values with the same parity as n (for example, if n is even then $\text{Prob}[S_t = \text{odd}] = 0$); e.g., for $n = 4$, $\text{Prob}[S_t = \pm 1] = \text{Prob}[S_t = \pm 3] = \text{Prob}[S_t = \pm 5] = 0$.

12.3 KOLMOGOROV–SMIRNOFF TESTS

A *one-sample Kolmogorov–Smirnoff* test is used to compare an observed cumulative distribution function (computed from a sample) to a specific continuous distribution function. This is a special test of goodness of fit. A *two-sample Kolmogorov–Smirnoff* test is used to compare two observed cumulative distribution functions; the null hypothesis is that the two independent samples come from identical continuous distributions.

12.3.1 One-sample Kolmogorov–Smirnoff test

Suppose a sample of size n is drawn from a population with known cumulative distribution function $F(x)$. The empirical distribution function, $F_n(x)$, is defined by the sample and is a step function given by

$$F_n(x) = \frac{k}{n} \quad \text{when} \quad x_{(i)} \leq x < x_{(i+1)} \tag{12.2}$$

where k is the number of observations less than or equal to x and the $\{x_{(i)}\}$ are the order statistics. If the sample is drawn from the hypothesized distribution, then the empirical distribution function, $F_n(x)$, should be close to

x	$F_n(x)$	$F(x) = x$	$|F_n(x) - F(x)|$
$x = 0$	0	0	$|0 - 0| = 0$
$x = .1-$	0	.10	$|0 - .10| = .10$
$x = .1+$	.25	.10	$|.25 - .10| = .15$
$x = .5-$	.25	.50	$|.25 - .50| = .25$
$x = .5+$	.50	.50	$|.50 - .50| = 0$
$x = .75-$	.50	.75	$|.50 - .75| = .25$
$x = .75+$	.75	.75	$|.75 - .75| = 0$
$x = .90-$	.75	.90	$|.75 - .90| = .15$
$x = .90+$	.90	.90	$|.90 - .90| = 0$
$x = 1$	1	1	$|1 - 1| = 0$

Table 12.1: Table for Kolmogorov–Smirnoff computation.

$F(x)$. Define the maximum absolute difference between the two distributions to be

$$D = \max \left| F_n(x) - F(x) \right| \qquad (12.3)$$

For a two-tailed test the table on page 170 gives critical values for the sampling distribution of D under the null hypothesis. One should reject the hypothetical distribution $F(x)$ if the value D exceeds the tabulated value.

A corresponding one-tailed test is provided by the statistic

$$D^+ = \max \left(F_n(x) - F(x) \right) \qquad (12.4)$$

Example 12.54: The values $\{.5, .75, .9, .1\}$ are observed from data that are presumed to be uniformly distributed on the interval $(0, 1)$. Since the presumed distribution is uniform, we have $F(x) = x$. Figure 12.1 shows $F_n(x)$ and $F(x)$. To determine D, only the values of $|F_n(x) - F(x)|$ for x at the endpoints ($x = 0$ and $x = 1$) and on each side of the sample values (since $F_n(x)$ has discontinuities at the sample values) need to be considered. Constructing Table 12.1 results in $D = .25$. If $\alpha = .05$ and $n = 4$ the table on page 170 yields a critical value of $c = .624$. Since $D < c$, the null hypothesis is not rejected.

12.3.2 Two-sample Kolmogorov–Smirnoff test

Suppose two independent samples of sizes n_1 and n_2 are drawn from a population with cumulative distribution function $F(x)$. For each sample j an empirical distribution function $F_{n_j}(x)$ is given by the step function,

$$F_{n_j}(x) = \frac{k}{n} \quad \text{when} \quad x_{(i)}^{\text{sample } j} \leq x < x_{(i+1)}^{\text{sample } j} \qquad (12.5)$$

where k is the number of observations less than or equal to x and the $\{x_{(i)}^{\text{sample } j}\}$ are the order statistics for the jth sample (for $j = 1$ or $j = 2$).

12.3. KOLMOGOROV–SMIRNOFF TESTS

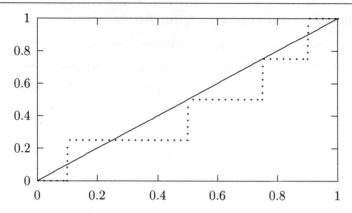

Figure 12.1: Comparison of the sample distribution function (dotted curve) with the distribution function (solid line) for a uniform random variable on the interval $(0,1)$.

If the two samples have been drawn from the same population, or from populations with the same distribution (the null hypothesis), then $F_{n_1}(x)$ should be close to $F_{n_2}(x)$. Define the maximum absolute difference between the two empirical distributions to be

$$D = \max \left| F_{n_1}(x) - F_{n_2}(x) \right| \qquad (12.6)$$

For a two-tailed test the table on page 172 gives critical values for the sampling distribution of D under the null hypothesis. The null hypothesis is rejected if the value of D exceeds the tabulated value.

A corresponding one-tailed test is provided by the statistic

$$D^+ = \max \left(F_{n_1}(x) - F_{n_2}(x) \right) \qquad (12.7)$$

12.3.3 Tables for Kolmogorov–Smirnoff tests

12.3.3.1 Critical values, one-sample Kolmogorov–Smirnoff test

Critical values, one-sample Kolmogorov–Smirnov test

One-sided test	$\alpha = 0.10$	0.05	0.025	0.01	0.005
Two-sided test	$\alpha = 0.20$	0.10	0.05	0.02	0.01
$n = 1$	0.900	0.950	0.975	0.990	0.995
2	0.684	0.776	0.842	0.900	0.929
3	0.565	0.636	0.708	0.785	0.829
4	0.493	0.565	0.624	0.689	0.734
5	0.447	0.509	0.563	0.627	0.669
6	0.410	0.468	0.519	0.577	0.617
7	0.381	0.436	0.483	0.538	0.576
8	0.358	0.410	0.454	0.507	0.542
9	0.339	0.387	0.430	0.480	0.513
10	0.323	0.369	0.409	0.457	0.489
11	0.308	0.352	0.391	0.437	0.468
12	0.296	0.338	0.375	0.419	0.449
13	0.285	0.325	0.361	0.404	0.432
14	0.275	0.314	0.349	0.390	0.418
15	0.266	0.304	0.338	0.377	0.404
16	0.258	0.295	0.327	0.366	0.392
17	0.250	0.286	0.318	0.355	0.381
18	0.244	0.279	0.309	0.346	0.371
19	0.237	0.271	0.301	0.337	0.361
20	0.232	0.265	0.294	0.329	0.352
21	0.226	0.259	0.287	0.321	0.344
22	0.221	0.253	0.281	0.314	0.337
23	0.216	0.247	0.275	0.307	0.330
24	0.212	0.242	0.269	0.301	0.323
25	0.208	0.238	0.264	0.295	0.317

Critical values, one-sample Kolmogorov–Smirnov test

One-sided test	$\alpha = 0.10$	0.05	0.025	0.01	0.005
Two-sided test	$\alpha = 0.20$	0.10	0.05	0.02	0.01
26	0.204	0.233	0.259	0.290	0.311
27	0.200	0.229	0.254	0.284	0.305
28	0.197	0.225	0.250	0.279	0.300
29	0.193	0.221	0.246	0.275	0.295
30	0.190	0.218	0.242	0.270	0.290
31	0.187	0.214	0.238	0.266	0.285
32	0.184	0.211	0.234	0.262	0.281
33	0.182	0.208	0.231	0.258	0.277
34	0.179	0.205	0.227	0.254	0.273
35	0.177	0.202	0.224	0.251	0.269
36	0.174	0.199	0.221	0.247	0.265
37	0.172	0.196	0.218	0.244	0.262
38	0.170	0.194	0.215	0.241	0.258
39	0.168	0.191	0.213	0.238	0.255
40	0.165	0.189	0.210	0.235	0.252
Approximation for $n > 40$:	$\dfrac{1.07}{\sqrt{n}}$	$\dfrac{1.22}{\sqrt{n}}$	$\dfrac{1.36}{\sqrt{n}}$	$\dfrac{1.52}{\sqrt{n}}$	$\dfrac{1.63}{\sqrt{n}}$

12.3.3.2 Critical values, two-sample Kolmogorov–Smirnoff test

Given the null hypothesis that the two distributions are the same (H_0: $F_1(x) = F_2(x)$), compute $D = \max |F_{n_1}(x) - F_{n_2}(x)|$.

(a) Reject H_0 if D exceeds the value in the table on page 172.

(b) Where $*$ appears in the table on page 172, do not reject H_0 at the given significance level.

(c) For large values of n_1 and n_2, and various values of α, the approximate critical value of D is given in the table below.

Level of significance	Approximate critical value
$\alpha = 0.10$	$1.22\sqrt{\dfrac{n_1+n_2}{n_1 n_2}}$
$\alpha = 0.05$	$1.36\sqrt{\dfrac{n_1+n_2}{n_1 n_2}}$
$\alpha = 0.025$	$1.48\sqrt{\dfrac{n_1+n_2}{n_1 n_2}}$
$\alpha = 0.01$	$1.63\sqrt{\dfrac{n_1+n_2}{n_1 n_2}}$
$\alpha = 0.005$	$1.73\sqrt{\dfrac{n_1+n_2}{n_1 n_2}}$
$\alpha = 0.001$	$1.95\sqrt{\dfrac{n_1+n_2}{n_1 n_2}}$

The entries in the following table are expressed as rational numbers since all critical values of D are an integer divided by $n_1 n_2$. For example, if $n_1 = 6$ and $n_2 = 5$, then

$$.108225 = \text{Prob}\left[D \geq \frac{20}{30}\right] = \text{Prob}\left[D \geq \frac{21}{30}\right] = \text{Prob}\left[D \geq \frac{22}{30}\right] = \text{Prob}\left[D \geq \frac{23}{30}\right]$$

$$.047619 = \text{Prob}\left[D \geq \frac{24}{30}\right] \quad \text{(least value of } D \text{ for which } \alpha < 0.05)$$

$$.025974 = \text{Prob}\left[D \geq \frac{25}{30}\right] = \text{Prob}\left[D \geq \frac{26}{30}\right] \quad (12.8)$$

$$= \text{Prob}\left[D \geq \frac{27}{30}\right] = \text{Prob}\left[D \geq \frac{28}{30}\right] = \text{Prob}\left[D \geq \frac{29}{30}\right]$$

$$.004329 = \text{Prob}\left[D \geq \frac{30}{30}\right] \quad \text{(least value of } D \text{ for which } \alpha < 0.01)$$

See P. J. Kim and R. I. Jennrich, Tables of the exact sampling distribution of the two-sample Kolmogorov–Smirnov criterion, D_{mn}, $m \leq n$, pages 79–170, in H. L. Harter and D. B. Brown (ed.), *Selected Tables in Mathematical Statistics*, Volume 1, American Mathematical Society, Providence, RI, 1973.

Critical values for the Kolmogorov–Smirnov test of $F_1(x) = F_2(x)$ (upper value for $\alpha \leq .05$, lower value for $\alpha \leq .01$)

Sample size n_2	Sample size n_1									
	3	4	5	6	7	8	9	10	11	12
1	*	*	*	*	*	*	*	*	*	*
	*	*	*	*	*	*	*	*	*	*
2	*	*	*	*	*	16/16	18/18	20/20	22/22	24/24
	*	*	*	*	*	*	*	*	*	*
3	*	*	15/15	18/18	21/21	21/24	24/27	27/30	30/33	30/36
	*	*	*	*	*	24/24	27/27	30/30	33/33	36/35
4		16/16	20/20	20/24	24/28	28/32	28/36	30/40	33/44	36/48
		*	*	24/24	28/28	32/32	32/36	36/40	40/44	44/48
5			*	24/30	30/35	30/40	35/45	40/50	39/55	43/60
			*	30/30	35/35	35/40	40/45	45/50	45/55	50/60
6				30/36	30/42	34/48	39/54	40/60	43/66	48/72
				36/36	36/42	40/48	45/54	48/60	54/66	60/72
7					42/49	40/56	42/63	46/70	48/77	53/84
					42/49	48/56	49/63	53/70	59/77	60/84
8						48/64	46/72	48/80	53/88	60/96
						56/64	55/72	60/80	64/88	68/96
9							54/81	53/90	59/99	63/108
							63/81	70/90	70/99	75/108
10								70/100	60/110	66/120
								80/100	77/110	80/120
11									77/121	72/132
									88/121	86/132
12										96/144
										84/144

12.4 KRUSKAL–WALLIS TEST

Assumptions: Suppose there are $k > 2$ independent random samples from continuous distributions, let n_i (for $i = 1, 2, \ldots, k$) be the number of observations in each sample, and let $n = n_1 + n_2 + \cdots + n_k$.

Hypothesis test:

H_0: the k samples are from identical populations.

H_a: at least two of the populations differ.

Rank all n observations from 1 (smallest) to n (largest). Equal observations are assigned the mean rank for their positions. Let R_{ij} be the rank assigned to the jth observation in the ith sample, and let R_i be the total of the ranks in the ith sample.

TS: $H = \left[\dfrac{12}{n(n+1)} \sum_{i=1}^{k} \dfrac{R_i^2}{n_i} \right] - 3(n+1)$

RR: $H \geq h$

where h is the critical value for the Kruskal–Wallis statistic (see table on page 174) such that $\text{Prob}\,[H \geq h] \approx \alpha$.

Note:

(1) The Kruskal–Wallis procedure is equivalent to an analysis of variance of the ranks. Define the variance ratio as

$$\text{VR} = \dfrac{\sum_{i=1}^{k} n_i \dfrac{\left(\overline{R}_i - \overline{\overline{R}}\right)^2}{k-1}}{\sum_{i=1}^{k} \sum_{j=1}^{n_i} \dfrac{\left(R_{ij} - \overline{R}_i\right)^2}{n-k}} \qquad (12.9)$$

where $\overline{R}_i = R_i/n_i$ is the mean of the ranks assigned to the ith sample and $\overline{\overline{R}} = (n+1)/2$ is the overall mean. The Kruskal–Wallis test statistic, H, and VR are related by the equations

$$\text{VR} = \dfrac{H(n-k)}{(k-1)(n-1-H)}, \qquad H = \dfrac{(n-1)(k-1)\text{VR}}{(n-k)+(k-1)\text{VR}}. \qquad (12.10)$$

(2) As $n \to \infty$ and each $n_i/n \to \lambda_i > 0$, H has approximately a chi-square distribution with $k-1$ degrees of freedom. Practically, if H_0 is true, and either

  (a) $k = 3$, $\quad n_i \geq 6$, $\quad i = 1, 2, 3$ or

  (b) $k > 3$, $\quad n_i \geq 5$, $\quad i = 1, 2, \ldots, k$

then H has a chi-square distribution with $k-1$ degrees of freedom.

(3) The variance ratio, VR, has approximately an F distribution with $k-1$ and $n-k$ degrees of freedom.

Example 12.55: Suppose that $k = 3$ treatments (A, B, and C) result in the following observations $\{1.2, 1.8, 1.7\}$, $\{0.9, 0.7\}$, and $\{1.0, 0.8\}$. (Therefore, $n_1 = 3$, $n_2 = 2$, $n_3 = 2$, $n = 7$.) Ranking these values:

Treatment	A	B	C
Sample size, n_i	3	2	2
Ranks	5	3	4
	7	1	2
	6		
Rank sums, R_i	18	4	6

Hence, $H = \frac{12}{7(8)}\left(\frac{18^2}{3} + \frac{4^2}{2} + \frac{6^2}{2}\right) - 3(8) = \frac{33}{7} \approx 4.714$. From the table below with $\{n_i\} = \{3, 2, 2\}$, we observe that Prob $[H \geq 4.714] = .0476$. At the $\alpha = .05$ level of significance, there is evidence to suggest at least two of the populations differ.

See R. L. Iman, D. Quade, and D. A. Alexander, Exact probability levels for the Kruskal–Wallis test, *Selected Tables in Mathematical Statistics*, Volume 3, American Mathematical Society, Providence, RI, 1975.

12.4.1 Tables for Kruskal–Wallis test

$\{n_i\} = \{2,1,1\}$		$\{n_i\} = \{2,2,1\}$		$\{n_i\} = \{2,2,2\}$		$\{n_i\} = \{3,2,1\}$	
h	$P(H \geq h)$	h	$P(H \geq h)$	h	$P(H \geq h)$	h	$P(H \geq h)$
2.700	0.5000	3.600	0.2000	4.571	0.0667	4.286	0.1000
				3.714	0.2000	3.857	0.1333

$\{n_i\} = \{3,2,2\}$		$\{n_i\} = \{3,3,1\}$		$\{n_i\} = \{3,3,2\}$		$\{n_i\} = \{3,3,3\}$	
h	$P(H \geq h)$	h	$P(H \geq h)$	h	$P(H \geq h)$	h	$P(H \geq h)$
5.357	0.0286	5.143	0.0429	6.250	0.0107	7.200	0.0036
4.714	0.0476	4.571	0.1000	5.556	0.0250	6.489	0.0107
4.500	0.0667	4.000	0.1286	5.361	0.0321	5.956	0.0250
4.464	0.1048	3.286	0.1571	5.139	0.0607	5.689	0.0286
3.929	0.1810	3.143	0.2429	5.000	0.0750	5.600	0.0500
3.750	0.2190	2.571	0.3286	4.694	0.0929	5.067	0.0857
3.607	0.2381	2.286	0.4857	4.556	0.1000	4.622	0.1000

$\{n_i\} = \{4,2,1\}$		$\{n_i\} = \{4,2,2\}$		$\{n_i\} = \{4,3,1\}$		$\{n_i\} = \{4,3,2\}$	
h	$P(H \geq h)$	h	$P(H \geq h)$	h	$P(H \geq h)$	h	$P(H \geq h)$
4.821	0.0571	6.000	0.0143	5.833	0.0214	7.000	0.0048
4.500	0.0762	5.500	0.0238	5.389	0.0357	6.444	0.0079
4.018	0.1143	5.333	0.0333	5.208	0.0500	6.300	0.0111
3.750	0.1333	5.125	0.0524	5.000	0.0571	6.111	0.0206
3.696	0.1714	4.500	0.0905	4.764	0.0714	5.800	0.0302
3.161	0.1905	4.458	0.1000	4.208	0.0786	5.500	0.0397
2.893	0.2667	4.167	0.1048	4.097	0.0857	5.400	0.0508
2.786	0.2857	4.125	0.1524	4.056	0.0929	4.444	0.1016

12.5 THE RUNS TEST

A **run** is a maximal subsequence of elements with a common property.

Hypothesis test:

H_0: the sequence is random.

H_a: the sequence is not random.

TS: $V = $ the total number of runs

RR: $V \geq v_1$ or $V \leq v_2$

where v_1 and v_2 are critical values for the runs test (see tables beginning on the next page) such that $\text{Prob}[V \geq v_1] \approx \alpha/2$ and $\text{Prob}[V \leq v_2] \approx \alpha/2$.

The normal approximation: Let m be the number of elements with the property that occurs least and n be the number of elements with the other property. As m and n increase, V has approximately a normal distribution with

$$\mu_V = \frac{2mn}{m+n} + 1 \quad \text{and} \quad \sigma_V^2 = \frac{2mn(2mn - m - n)}{(m+n)^2(m+n+1)}. \tag{12.11}$$

The random variable

$$Z = \frac{V - \mu_V}{\sigma_V} \tag{12.12}$$

has approximately a standard normal distribution.

Example 12.56: Suppose the following sequence of heads (H) and tails (T) was obtained from flipping a coin: $\{H, H, T, T, H, T, H, T, T, T, T, H\}$. Is there any evidence to suggest the coin is biased?

Solution:

(S1) Place vertical bars at the end of each run. The data set may be written to easily count the number of runs.

$$HH \mid TT \mid H \mid T \mid H \mid TTTT \mid H \mid$$

(S2) Using this notation, there are 5 H's, 7 T's, and 7 runs.

(S3) The table on page 177 (using $m = 5$ and $n = 7$) indicates that 65% of the time one would expect there to be 7 runs or fewer.

(S4) The table on page 177 (using $m = 5$ and $n = 6$) indicates that 42% of the time one would expect there be 6 runs or fewer. Alternatively, 58% (since $1 - 0.42 = 0.58$) of the time there would be 7 runs or more.

(S5) In neither case is there any evidence to suggest the coin is biased.

12.5.1 Tables for the runs test

The following tables give the sampling distribution for v for values of m and n less than or equal to 10. That is, the values listed in this table give the probability that v or fewer runs will occur.

The table on page 184 gives percentage points of the distribution for larger sample sizes when $m = n$. The columns headed with 0.5%, 1%, 2.5%, 5% indicate the values of v such that v or fewer runs occur with probability less than that indicated; the columns headed with 97.5%, 99%, 99.5% indicate values of v for which the probability of v or more runs is less than 2.5%, 1%, 0.5%. For large values of m and n, particularly for $m = n$ greater than 10, a normal approximation may be used, with the parameters given in equation (12.11).

Distribution of total number of runs v in samples of size (m, n)

m, n	$v = 2$	3	4	5	6	7	8	9	10
2, 2	0.3333	0.6667	1.0000						
2, 3	0.2000	0.5000	0.9000	1.0000					
2, 4	0.1333	0.4000	0.8000	1.0000					
2, 5	0.0952	0.3333	0.7143	1.0000					
2, 6	0.0714	0.2857	0.6429	1.0000					
2, 7	0.0556	0.2500	0.5833	1.0000					
2, 8	0.0444	0.2222	0.5333	1.0000					
2, 9	0.0364	0.2000	0.4909	1.0000					
2, 10	0.0303	0.1818	0.4545	1.0000					
2, 11	0.0256	0.1667	0.4231	1.0000					
2, 12	0.0220	0.1538	0.3956	1.0000					
2, 13	0.0190	0.1429	0.3714	1.0000					
2, 14	0.0167	0.1333	0.3500	1.0000					
2, 15	0.0147	0.1250	0.3309	1.0000					
2, 16	0.0131	0.1176	0.3137	1.0000					
2, 17	0.0117	0.1111	0.2982	1.0000					
2, 18	0.0105	0.1053	0.2842	1.0000					
2, 19	0.0095	0.1000	0.2714	1.0000					
2, 20	0.0087	0.0952	0.2597	1.0000					
3, 3	0.1000	0.3000	0.7000	0.9000	1.0000				
3, 4	0.0571	0.2000	0.5429	0.8000	0.9714	1.0000			
3, 5	0.0357	0.1429	0.4286	0.7143	0.9286	1.0000			
3, 6	0.0238	0.1071	0.3452	0.6429	0.8810	1.0000			
3, 7	0.0167	0.0833	0.2833	0.5833	0.8333	1.0000			
3, 8	0.0121	0.0667	0.2364	0.5333	0.7879	1.0000			
3, 9	0.0091	0.0545	0.2000	0.4909	0.7455	1.0000			
3, 10	0.0070	0.0455	0.1713	0.4545	0.7063	1.0000			
3, 11	0.0055	0.0385	0.1484	0.4231	0.6703	1.0000			
3, 12	0.0044	0.0330	0.1297	0.3956	0.6374	1.0000			
3, 13	0.0036	0.0286	0.1143	0.3714	0.6071	1.0000			
3, 14	0.0029	0.0250	0.1015	0.3500	0.5794	1.0000			
3, 15	0.0025	0.0221	0.0907	0.3309	0.5539	1.0000			
3, 16	0.0021	0.0196	0.0815	0.3137	0.5304	1.0000			
3, 17	0.0018	0.0175	0.0737	0.2982	0.5088	1.0000			
3, 18	0.0015	0.0158	0.0669	0.2842	0.4887	1.0000			
3, 19	0.0013	0.0143	0.0610	0.2714	0.4701	1.0000			
3, 20	0.0011	0.0130	0.0559	0.2597	0.4529	1.0000			

12.5. THE RUNS TEST

Distribution of total number of runs v in samples of size (m, n)

m, n	$v = 2$	3	4	5	6	7	8	9	10
4, 4	0.0286	0.1143	0.3714	0.6286	0.8857	0.9714	1.0000		
4, 5	0.0159	0.0714	0.2619	0.5000	0.7857	0.9286	0.9921	1.0000	
4, 6	0.0095	0.0476	0.1905	0.4048	0.6905	0.8810	0.9762	1.0000	
4, 7	0.0061	0.0333	0.1424	0.3333	0.6061	0.8333	0.9545	1.0000	
4, 8	0.0040	0.0242	0.1091	0.2788	0.5333	0.7879	0.9293	1.0000	
4, 9	0.0028	0.0182	0.0853	0.2364	0.4713	0.7455	0.9021	1.0000	
4, 10	0.0020	0.0140	0.0679	0.2028	0.4186	0.7063	0.8741	1.0000	
4, 11	0.0015	0.0110	0.0549	0.1758	0.3736	0.6703	0.8462	1.0000	
4, 12	0.0011	0.0088	0.0451	0.1538	0.3352	0.6374	0.8187	1.0000	
4, 13	0.0008	0.0071	0.0374	0.1357	0.3021	0.6071	0.7920	1.0000	
4, 14	0.0007	0.0059	0.0314	0.1206	0.2735	0.5794	0.7663	1.0000	
4, 15	0.0005	0.0049	0.0266	0.1078	0.2487	0.5539	0.7417	1.0000	
4, 16	0.0004	0.0041	0.0227	0.0970	0.2270	0.5304	0.7183	1.0000	
4, 17	0.0003	0.0035	0.0195	0.0877	0.2080	0.5088	0.6959	1.0000	
4, 18	0.0003	0.0030	0.0170	0.0797	0.1913	0.4887	0.6746	1.0000	
4, 19	0.0002	0.0026	0.0148	0.0727	0.1764	0.4701	0.6544	1.0000	
4, 20	0.0002	0.0023	0.0130	0.0666	0.1632	0.4529	0.6352	1.0000	
5, 5	0.0079	0.0397	0.1667	0.3571	0.6429	0.8333	0.9603	0.9921	1.0000
5, 6	0.0043	0.0238	0.1104	0.2619	0.5216	0.7381	0.9113	0.9762	0.9978
5, 7	0.0025	0.0152	0.0758	0.1970	0.4242	0.6515	0.8535	0.9545	0.9924
5, 8	0.0016	0.0101	0.0536	0.1515	0.3473	0.5758	0.7933	0.9293	0.9837
5, 9	0.0010	0.0070	0.0390	0.1189	0.2867	0.5105	0.7343	0.9021	0.9720
5, 10	0.0007	0.0050	0.0290	0.0949	0.2388	0.4545	0.6783	0.8741	0.9580
5, 11	0.0005	0.0037	0.0220	0.0769	0.2005	0.4066	0.6264	0.8462	0.9423
5, 12	0.0003	0.0027	0.0170	0.0632	0.1698	0.3654	0.5787	0.8187	0.9253
5, 13	0.0002	0.0021	0.0133	0.0525	0.1450	0.3298	0.5352	0.7920	0.9076
5, 14	0.0002	0.0016	0.0106	0.0441	0.1246	0.2990	0.4958	0.7663	0.8893
5, 15	0.0001	0.0013	0.0085	0.0374	0.1078	0.2722	0.4600	0.7417	0.8709
5, 16	$.0^4 983$	0.0010	0.0069	0.0320	0.0939	0.2487	0.4276	0.7183	0.8524
5, 17	$.0^4 759$	0.0008	0.0057	0.0276	0.0823	0.2281	0.3982	0.6959	0.8341
5, 18	$.0^4 594$	0.0007	0.0047	0.0239	0.0724	0.2098	0.3715	0.6746	0.8161
5, 19	$.0^4 471$	0.0006	0.0040	0.0209	0.0641	0.1937	0.3473	0.6544	0.7984
5, 20	$.0^4 376$	0.0005	0.0033	0.0184	0.0570	0.1793	0.3252	0.6352	0.7811
6, 6	0.0022	0.0130	0.0671	0.1753	0.3918	0.6082	0.8247	0.9329	0.9870
6, 7	0.0012	0.0076	0.0425	0.1212	0.2960	0.5000	0.7331	0.8788	0.9662
6, 8	0.0007	0.0047	0.0280	0.0862	0.2261	0.4126	0.6457	0.8205	0.9371
6, 9	0.0004	0.0030	0.0190	0.0629	0.1748	0.3427	0.5664	0.7622	0.9021
6, 10	0.0002	0.0020	0.0132	0.0470	0.1369	0.2867	0.4965	0.7063	0.8636
6, 11	0.0002	0.0014	0.0095	0.0357	0.1084	0.2418	0.4357	0.6538	0.8235
6, 12	0.0001	0.0010	0.0069	0.0276	0.0869	0.2054	0.3832	0.6054	0.7831
6, 13	$.0^4 737$	0.0007	0.0051	0.0217	0.0704	0.1758	0.3379	0.5609	0.7434
6, 14	$.0^4 516$	0.0005	0.0039	0.0173	0.0575	0.1514	0.2990	0.5204	0.7049
6, 15	$.0^4 369$	0.0004	0.0030	0.0139	0.0475	0.1313	0.2655	0.4835	0.6680
6, 16	$.0^4 268$	0.0003	0.0023	0.0114	0.0395	0.1146	0.2365	0.4499	0.6329
6, 17	$.0^4 198$	0.0002	0.0018	0.0093	0.0331	0.1005	0.2114	0.4195	0.5998
6, 18	$.0^4 149$	0.0002	0.0014	0.0078	0.0280	0.0886	0.1896	0.3917	0.5685
6, 19	$.0^4 113$	0.0001	0.0012	0.0065	0.0238	0.0785	0.1706	0.3665	0.5392
6, 20	$.0^5 869$	0.0001	0.0009	0.0055	0.0203	0.0698	0.1540	0.3434	0.5118

Distribution of total number of runs v in samples of size (m, n)

m, n	$v = 11$	12	13	14	15	16	17	18	19	20	21
5, 5											
5, 6	1.0000										
5, 7	1.0000										
5, 8	1.0000										
5, 9	1.0000										
5, 10	1.0000										
5, 11	1.0000										
5, 12	1.0000										
5, 13	1.0000										
5, 14	1.0000										
5, 15	1.0000										
5, 16	1.0000										
5, 17	1.0000										
5, 18	1.0000										
5, 19	1.0000										
5, 20	1.0000										
6, 6	0.9978	1.0000									
6, 7	0.9924	0.9994	1.0000								
6, 8	0.9837	0.9977	1.0000								
6, 9	0.9720	0.9944	1.0000								
6, 10	0.9580	0.9895	1.0000								
6, 11	0.9423	0.9830	1.0000								
6, 12	0.9253	0.9751	1.0000								
6, 13	0.9076	0.9659	1.0000								
6, 14	0.8893	0.9557	1.0000								
6, 15	0.8709	0.9447	1.0000								
6, 16	0.8524	0.9329	1.0000								
6, 17	0.8341	0.9207	1.0000								
6, 18	0.8161	0.9081	1.0000								
6, 19	0.7984	0.8952	1.0000								
6, 20	0.7811	0.8822	1.0000								

12.5. THE RUNS TEST

Distribution of total number of runs v in samples of size (m, n)

m, n	$v = 2$	3	4	5	6	7	8	9	10
7, 7	0.0006	0.0041	0.0251	0.0775	0.2086	0.3834	0.6166	0.7914	0.9225
7, 8	0.0003	0.0023	0.0154	0.0513	0.1492	0.2960	0.5136	0.7040	0.8671
7, 9	0.0002	0.0014	0.0098	0.0350	0.1084	0.2308	0.4266	0.6224	0.8059
7, 10	0.0001	0.0009	0.0064	0.0245	0.0800	0.1818	0.3546	0.5490	0.7433
7, 11	$.0^4 628$	0.0006	0.0043	0.0175	0.0600	0.1448	0.2956	0.4842	0.6821
7, 12	$.0^4 397$	0.0004	0.0030	0.0128	0.0456	0.1165	0.2475	0.4276	0.6241
7, 13	$.0^4 258$	0.0003	0.0021	0.0095	0.0351	0.0947	0.2082	0.3785	0.5700
7, 14	$.0^4 172$	0.0002	0.0015	0.0072	0.0273	0.0777	0.1760	0.3359	0.5204
7, 15	$.0^4 117$	0.0001	0.0011	0.0055	0.0216	0.0642	0.1496	0.2990	0.4751
7, 16	$.0^5 816$	$.0^4 938$	0.0008	0.0043	0.0172	0.0536	0.1278	0.2670	0.4340
7, 17	$.0^5 578$	$.0^4 693$	0.0006	0.0034	0.0138	0.0450	0.1097	0.2392	0.3969
7, 18	$.0^5 416$	$.0^4 520$	0.0005	0.0027	0.0112	0.0381	0.0947	0.2149	0.3634
7, 19	$.0^5 304$	$.0^4 395$	0.0004	0.0022	0.0092	0.0324	0.0820	0.1937	0.3332
7, 20	$.0^5 225$	$.0^4 304$	0.0003	0.0018	0.0075	0.0278	0.0714	0.1751	0.3060
8, 8	0.0002	0.0012	0.0089	0.0317	0.1002	0.2145	0.4048	0.5952	0.7855
8, 9	$.0^4 823$	0.0007	0.0053	0.0203	0.0687	0.1573	0.3186	0.5000	0.7016
8, 10	$.0^4 457$	0.0004	0.0033	0.0134	0.0479	0.1170	0.2514	0.4194	0.6209
8, 11	$.0^4 265$	0.0003	0.0021	0.0090	0.0341	0.0882	0.1994	0.3522	0.5467
8, 12	$.0^4 159$	0.0002	0.0014	0.0063	0.0246	0.0674	0.1591	0.2966	0.4800
8, 13	$.0^5 983$	0.0001	0.0009	0.0044	0.0181	0.0521	0.1278	0.2508	0.4211
8, 14	$.0^5 625$	$.0^4 688$	0.0006	0.0032	0.0134	0.0408	0.1034	0.2129	0.3695
8, 15	$.0^5 408$	$.0^4 469$	0.0004	0.0023	0.0101	0.0322	0.0842	0.1816	0.3245
8, 16	$.0^5 272$	$.0^4 326$	0.0003	0.0017	0.0077	0.0257	0.0690	0.1556	0.2856
8, 17	$.0^5 185$	$.0^4 231$	0.0002	0.0013	0.0060	0.0207	0.0570	0.1340	0.2518
8, 18	$.0^5 128$	$.0^4 166$	0.0002	0.0010	0.0047	0.0169	0.0473	0.1159	0.2225
8, 19	$.0^6 901$	$.0^4 122$	0.0001	0.0008	0.0037	0.0138	0.0395	0.1006	0.1971
8, 20	$.0^6 643$	$.0^5 901$	$.0^4 946$	0.0006	0.0029	0.0114	0.0332	0.0878	0.1751
9, 9	$.0^4 411$	0.0004	0.0030	0.0122	0.0445	0.1090	0.2380	0.3992	0.6008
9, 10	$.0^4 217$	0.0002	0.0018	0.0076	0.0294	0.0767	0.1786	0.3186	0.5095
9, 11	$.0^4 119$	0.0001	0.0011	0.0049	0.0199	0.0549	0.1349	0.2549	0.4300
9, 12	$.0^5 680$	$.0^4 714$	0.0007	0.0032	0.0137	0.0399	0.1028	0.2049	0.3621
9, 13	$.0^5 402$	$.0^4 442$	0.0004	0.0022	0.0096	0.0294	0.0789	0.1656	0.3050
9, 14	$.0^5 245$	$.0^4 281$	0.0003	0.0015	0.0068	0.0220	0.0612	0.1347	0.2572
9, 15	$.0^5 153$	$.0^4 184$	0.0002	0.0010	0.0049	0.0166	0.0478	0.1102	0.2174
9, 16	$.0^6 979$	$.0^4 122$	0.0001	0.0007	0.0036	0.0127	0.0377	0.0907	0.1842
9, 17	$.0^6 640$	$.0^5 832$	$.0^4 903$	0.0005	0.0027	0.0099	0.0299	0.0751	0.1566
9, 18	$.0^6 427$	$.0^5 576$	$.0^4 638$	0.0004	0.0020	0.0077	0.0240	0.0626	0.1336
9, 19	$.0^6 290$	$.0^5 405$	$.0^4 458$	0.0003	0.0015	0.0061	0.0193	0.0524	0.1144
9, 20	$.0^6 200$	$.0^5 290$	$.0^4 333$	0.0002	0.0012	0.0048	0.0157	0.0441	0.0983
10, 10	$.0^4 108$	0.0001	0.0010	0.0045	0.0185	0.0513	0.1276	0.2422	0.4141
10, 11	$.0^5 567$	$.0^4 595$	0.0006	0.0027	0.0119	0.0349	0.0920	0.1849	0.3350
10, 12	$.0^5 309$	$.0^4 340$	0.0003	0.0017	0.0078	0.0242	0.0670	0.1421	0.2707
10, 13	$.0^5 175$	$.0^4 201$	0.0002	0.0011	0.0053	0.0170	0.0493	0.1099	0.2189
10, 14	$.0^5 102$	$.0^4 122$	0.0001	0.0007	0.0036	0.0122	0.0367	0.0857	0.1775
10, 15	$.0^6 612$	$.0^5 765$	$.0^4 847$	0.0005	0.0025	0.0088	0.0275	0.0673	0.1445
10, 16	$.0^6 377$	$.0^5 489$	$.0^4 557$	0.0003	0.0018	0.0065	0.0209	0.0533	0.1180
10, 17	$.0^6 237$	$.0^5 320$	$.0^4 373$	0.0002	0.0013	0.0048	0.0160	0.0425	0.0968
10, 18	$.0^6 152$	$.0^5 213$	$.0^4 255$	0.0002	0.0009	0.0036	0.0124	0.0341	0.0798
10, 19	$.0^7 999$	$.0^5 145$	$.0^4 176$	0.0001	0.0007	0.0028	0.0096	0.0276	0.0661
10, 20	$.0^7 666$	$.0^6 999$	$.0^4 124$	$.0^4 864$	0.0005	0.0021	0.0076	0.0225	0.0550

Distribution of total number of runs v in samples of size (m, n)

m, n	$v = 11$	12	13	14	15	16	17	18	19	20	21
7, 7	0.9749	0.9959	0.9994	1.0000							
7, 8	0.9487	0.9879	0.9977	0.9998	1.0000						
7, 9	0.9161	0.9748	0.9944	0.9993	1.0000						
7, 10	0.8794	0.9571	0.9895	0.9981	1.0000						
7, 11	0.8405	0.9355	0.9830	0.9962	1.0000						
7, 12	0.8009	0.9109	0.9751	0.9935	1.0000						
7, 13	0.7616	0.8842	0.9659	0.9898	1.0000						
7, 14	0.7233	0.8561	0.9557	0.9852	1.0000						
7, 15	0.6864	0.8273	0.9447	0.9799	1.0000						
7, 16	0.6512	0.7982	0.9329	0.9738	1.0000						
7, 17	0.6178	0.7692	0.9207	0.9669	1.0000						
7, 18	0.5862	0.7407	0.9081	0.9595	1.0000						
7, 19	0.5565	0.7128	0.8952	0.9516	1.0000						
7, 20	0.5286	0.6857	0.8822	0.9433	1.0000						
8, 8	0.8998	0.9683	0.9911	0.9988	0.9998	1.0000					
8, 9	0.8427	0.9394	0.9797	0.9958	0.9993	1.0000	1.0000				
8, 10	0.7822	0.9031	0.9636	0.9905	0.9981	0.9998	1.0000				
8, 11	0.7217	0.8618	0.9434	0.9823	0.9962	0.9994	1.0000				
8, 12	0.6634	0.8174	0.9201	0.9714	0.9935	0.9987	1.0000				
8, 13	0.6084	0.7718	0.8944	0.9580	0.9898	0.9976	1.0000				
8, 14	0.5573	0.7263	0.8672	0.9423	0.9852	0.9960	1.0000				
8, 15	0.5103	0.6818	0.8390	0.9248	0.9799	0.9939	1.0000				
8, 16	0.4674	0.6389	0.8104	0.9057	0.9738	0.9913	1.0000				
8, 17	0.4285	0.5981	0.7818	0.8855	0.9669	0.9881	1.0000				
8, 18	0.3931	0.5595	0.7536	0.8645	0.9595	0.9844	1.0000				
8, 19	0.3611	0.5232	0.7258	0.8429	0.9516	0.9803	1.0000				
8, 20	0.3322	0.4893	0.6988	0.8210	0.9433	0.9757	1.0000				
9, 9	0.7620	0.8910	0.9555	0.9878	0.9970	0.9996	1.0000	1.0000			
9, 10	0.6814	0.8342	0.9233	0.9742	0.9924	0.9986	0.9998	1.0000	1.0000		
9, 11	0.6050	0.7731	0.8851	0.9551	0.9851	0.9965	0.9994	0.9999	1.0000		
9, 12	0.5350	0.7111	0.8431	0.9311	0.9751	0.9931	0.9987	0.9998	1.0000		
9, 13	0.4721	0.6505	0.7991	0.9031	0.9625	0.9880	0.9976	0.9996	1.0000		
9, 14	0.4164	0.5928	0.7545	0.8721	0.9477	0.9813	0.9960	0.9991	1.0000		
9, 15	0.3674	0.5389	0.7104	0.8390	0.9309	0.9729	0.9939	0.9985	1.0000		
9, 16	0.3245	0.4892	0.6675	0.8047	0.9125	0.9629	0.9913	0.9976	1.0000		
9, 17	0.2871	0.4437	0.6264	0.7699	0.8929	0.9515	0.9881	0.9963	1.0000		
9, 18	0.2545	0.4024	0.5872	0.7351	0.8724	0.9388	0.9844	0.9948	1.0000		
9, 19	0.2261	0.3650	0.5503	0.7008	0.8513	0.9250	0.9803	0.9930	1.0000		
9, 20	0.2013	0.3313	0.5155	0.6672	0.8298	0.9103	0.9757	0.9908	1.0000		
10, 10	0.5859	0.7578	0.8724	0.9487	0.9815	0.9955	0.9990	0.9999	1.0000	1.0000	
10, 11	0.5000	0.6800	0.8151	0.9151	0.9651	0.9896	0.9973	0.9996	0.9999	1.0000	
10, 12	0.4250	0.6050	0.7551	0.8751	0.9437	0.9804	0.9942	0.9988	0.9998	1.0000	1.0000
10, 13	0.3607	0.5351	0.6950	0.8307	0.9180	0.9678	0.9896	0.9974	0.9996	0.9999	1.0000
10, 14	0.3062	0.4715	0.6369	0.7839	0.8889	0.9519	0.9834	0.9952	0.9991	0.9999	1.0000
10, 15	0.2602	0.4146	0.5818	0.7361	0.8574	0.9330	0.9755	0.9920	0.9985	0.9997	1.0000
10, 16	0.2216	0.3641	0.5303	0.6886	0.8243	0.9115	0.9661	0.9879	0.9976	0.9994	1.0000
10, 17	0.1893	0.3197	0.4828	0.6423	0.7904	0.8880	0.9551	0.9826	0.9963	0.9991	1.0000
10, 18	0.1621	0.2809	0.4393	0.5978	0.7562	0.8629	0.9429	0.9763	0.9948	0.9985	1.0000
10, 19	0.1392	0.2470	0.3997	0.5554	0.7223	0.8367	0.9296	0.9689	0.9930	0.9978	1.0000
10, 20	0.1200	0.2175	0.3638	0.5155	0.6889	0.8097	0.9153	0.9606	0.9908	0.9969	1.0000

Distribution of total number of runs v in samples of size (m, n)

m, n	$v = 2$	3	4	5	6	7	8	9	10
11, 11	$.0^5 284$	$.0^4 312$	0.0003	0.0016	0.0073	0.0226	0.0635	0.1349	0.2599
11, 12	$.0^5 148$	$.0^4 170$	0.0002	0.0010	0.0046	0.0150	0.0443	0.0992	0.2017
11, 13	$.0^6 801$	$.0^5 961$	0.0001	0.0006	0.0030	0.0101	0.0313	0.0736	0.1569
11, 14	$.0^6 449$	$.0^5 561$	$.0^4 639$	0.0004	0.0019	0.0069	0.0223	0.0551	0.1224
11, 15	$.0^6 259$	$.0^5 337$	$.0^4 396$	0.0002	0.0013	0.0048	0.0161	0.0416	0.0960
11, 16	$.0^6 153$	$.0^5 207$	$.0^4 251$	0.0002	0.0009	0.0034	0.0118	0.0317	0.0757
11, 17	$.0^7 931$	$.0^5 130$	$.0^4 162$	0.0001	0.0006	0.0025	0.0087	0.0244	0.0600
11, 18	$.0^7 578$	$.0^6 838$	$.0^4 107$	$.0^4 721$	0.0004	0.0018	0.0065	0.0189	0.0478
11, 19	$.0^7 366$	$.0^6 549$	$.0^5 714$	$.0^4 500$	0.0003	0.0013	0.0049	0.0148	0.0383
11, 20	$.0^7 236$	$.0^6 366$	$.0^5 485$	$.0^4 351$	0.0002	0.0010	0.0037	0.0116	0.0308
12, 12	$.0^6 740$	$.0^5 888$	$.0^4 984$	0.0005	0.0028	0.0095	0.0296	0.0699	0.1504
12, 13	$.0^6 385$	$.0^5 481$	$.0^4 556$	0.0003	0.0017	0.0061	0.0201	0.0498	0.1126
12, 14	$.0^6 207$	$.0^5 269$	$.0^4 323$	0.0002	0.0011	0.0040	0.0138	0.0358	0.0847
12, 15	$.0^6 115$	$.0^5 155$	$.0^4 193$	0.0001	0.0007	0.0027	0.0096	0.0260	0.0640
12, 16	$.0^7 657$	$.0^6 920$	$.0^4 118$	$.0^4 769$	0.0005	0.0018	0.0068	0.0191	0.0487
12, 17	$.0^7 385$	$.0^6 559$	$.0^5 734$	$.0^4 497$	0.0003	0.0013	0.0048	0.0142	0.0373
12, 18	$.0^7 231$	$.0^6 347$	$.0^5 467$	$.0^4 328$	0.0002	0.0009	0.0035	0.0106	0.0288
12, 19	$.0^7 142$	$.0^6 220$	$.0^5 303$	$.0^4 220$	0.0001	0.0006	0.0025	0.0080	0.0223
12, 20	$.0^8 886$	$.0^6 142$	$.0^5 199$	$.0^4 150$	$.0^4 983$	0.0005	0.0019	0.0061	0.0175
13, 13	$.0^6 192$	$.0^5 250$	$.0^4 302$	0.0002	0.0010	0.0038	0.0131	0.0341	0.0812
13, 14	$.0^7 997$	$.0^5 135$	$.0^4 169$	0.0001	0.0006	0.0024	0.0087	0.0236	0.0589
13, 15	$.0^7 534$	$.0^6 748$	$.0^5 972$	$.0^4 636$	0.0004	0.0016	0.0058	0.0165	0.0430
13, 16	$.0^7 295$	$.0^6 427$	$.0^5 573$	$.0^4 389$	0.0002	0.0010	0.0040	0.0117	0.0316
13, 17	$.0^7 167$	$.0^6 251$	$.0^5 346$	$.0^4 243$	0.0002	0.0007	0.0027	0.0084	0.0234
13, 18	$.0^8 970$	$.0^6 150$	$.0^5 213$	$.0^4 155$	0.0001	0.0005	0.0019	0.0061	0.0175
13, 19	$.0^8 576$	$.0^7 921$	$.0^5 134$	$.0^4 100$	$.0^4 682$	0.0003	0.0014	0.0045	0.0132
13, 20	$.0^8 349$	$.0^7 576$	$.0^6 853$	$.0^5 662$	$.0^4 460$	0.0002	0.0010	0.0033	0.0100
14, 14	$.0^7 499$	$.0^6 698$	$.0^5 912$	$.0^4 597$	0.0004	0.0015	0.0056	0.0157	0.0412
14, 15	$.0^7 258$	$.0^6 374$	$.0^5 507$	$.0^4 344$	0.0002	0.0009	0.0036	0.0107	0.0291
14, 16	$.0^7 138$	$.0^6 206$	$.0^5 289$	$.0^4 203$	0.0001	0.0006	0.0024	0.0073	0.0207
14, 17	$.0^8 754$	$.0^6 117$	$.0^5 169$	$.0^4 123$	$.0^4 829$	0.0004	0.0016	0.0051	0.0149
14, 18	$.0^8 424$	$.0^7 679$	$.0^5 101$	$.0^5 757$	$.0^4 526$	0.0002	0.0011	0.0035	0.0108
14, 19	$.0^8 244$	$.0^7 403$	$.0^6 612$	$.0^5 476$	$.0^4 339$	0.0002	0.0007	0.0025	0.0079
14, 20	$.0^8 144$	$.0^7 244$	$.0^6 379$	$.0^5 304$	$.0^4 222$	0.0001	0.0005	0.0018	0.0058
15, 15	$.0^7 129$	$.0^6 193$	$.0^5 272$	$.0^4 191$	0.0001	0.0006	0.0023	0.0070	0.0199
15, 16	$.0^8 665$	$.0^6 103$	$.0^5 150$	$.0^4 109$	$.0^4 745$	0.0003	0.0014	0.0046	0.0137
15, 17	$.0^8 354$	$.0^7 566$	$.0^6 848$	$.0^5 639$	$.0^4 450$	0.0002	0.0009	0.0031	0.0095
15, 18	$.0^8 193$	$.0^7 318$	$.0^6 491$	$.0^5 382$	$.0^4 277$	0.0001	0.0006	0.0021	0.0067
15, 19	$.0^8 108$	$.0^7 183$	$.0^6 290$	$.0^5 233$	$.0^4 173$	$.0^4 873$	0.0004	0.0014	0.0047
15, 20	$.0^9 616$	$.0^7 108$	$.0^6 175$	$.0^5 144$	$.0^4 110$	$.0^4 573$	0.0003	0.0010	0.0034
16, 16	$.0^8 333$	$.0^7 532$	$.0^6 802$	$.0^5 604$	$.0^4 427$	0.0002	0.0009	0.0030	0.0092
16, 17	$.0^8 171$	$.0^7 283$	$.0^6 440$	$.0^5 342$	$.0^4 250$	0.0001	0.0006	0.0019	0.0062
16, 18	$.0^9 907$	$.0^7 154$	$.0^6 247$	$.0^5 198$	$.0^4 149$	$.0^4 754$	0.0004	0.0013	0.0042
16, 19	$.0^9 493$	$.0^8 862$	$.0^6 142$	$.0^5 117$	$.0^5 909$	$.0^4 473$	0.0002	0.0008	0.0029
16, 20	$.0^9 274$	$.0^8 493$	$.0^7 829$	$.0^6 707$	$.0^5 562$	$.0^4 302$	0.0002	0.0006	0.0020
17, 17	$.0^9 857$	$.0^7 146$	$.0^6 234$	$.0^5 188$	$.0^4 142$	$.0^4 718$	0.0003	0.0012	0.0041
17, 18	$.0^9 441$	$.0^8 771$	$.0^6 128$	$.0^5 106$	$.0^5 825$	$.0^4 430$	0.0002	0.0008	0.0027
17, 19	$.0^9 233$	$.0^8 419$	$.0^7 712$	$.0^6 607$	$.0^5 488$	$.0^4 262$	0.0001	0.0005	0.0018
17, 20	$.0^9 126$	$.0^8 233$	$.0^7 406$	$.0^6 356$	$.0^5 294$	$.0^4 163$	$.0^4 845$	0.0003	0.0012
18, 18	$.0^9 220$	$.0^8 397$	$.0^7 677$	$.0^6 577$	$.0^5 465$	$.0^4 250$	0.0001	0.0005	0.0017
18, 19	$.0^9 113$	$.0^8 209$	$.0^7 367$	$.0^6 322$	$.0^5 268$	$.0^4 148$	$.0^4 776$	0.0003	0.0011
18, 20	$.0^{10} 596$	$.0^8 113$	$.0^7 204$	$.0^6 184$	$.0^5 157$	$.0^5 896$	$.0^4 482$	0.0002	0.0007
19, 19	$.0^{10} 566$	$.0^8 108$	$.0^7 194$	$.0^6 175$	$.0^5 150$	$.0^5 856$	$.0^4 462$	0.0002	0.0007
19, 20	$.0^{10} 290$	$.0^9 566$	$.0^7 105$	$.0^7 973$	$.0^6 857$	$.0^5 503$	$.0^4 280$	0.0001	0.0005
20, 20	$.0^{10} 145$	$.0^9 290$	$.0^8 553$	$.0^7 527$	$.0^6 477$	$.0^5 288$	$.0^4 165$	$.0^4 710$	0.0003

Distribution of total number of runs v in samples of size (m, n)

m, n	$v = 11$	12	13	14	15	16	17	18	19	20	21
11, 11	0.4100	0.5900	0.7401	0.8651	0.9365	0.9774	0.9927	0.9984	0.9997	1.0000	1.0000
11, 12	0.3350	0.5072	0.6650	0.8086	0.9008	0.9594	0.9850	0.9960	0.9990	0.9999	1.0000
11, 13	0.2735	0.4334	0.5933	0.7488	0.8598	0.9360	0.9740	0.9919	0.9978	0.9996	0.9999
11, 14	0.2235	0.3690	0.5267	0.6883	0.8154	0.9078	0.9598	0.9857	0.9958	0.9991	0.9999
11, 15	0.1831	0.3137	0.4660	0.6293	0.7692	0.8758	0.9424	0.9774	0.9930	0.9981	0.9997
11, 16	0.1504	0.2665	0.4116	0.5728	0.7225	0.8410	0.9224	0.9669	0.9891	0.9967	0.9994
11, 17	0.1240	0.2265	0.3632	0.5199	0.6765	0.8043	0.9002	0.9542	0.9841	0.9948	0.9991
11, 18	0.1027	0.1928	0.3205	0.4708	0.6317	0.7666	0.8763	0.9395	0.9781	0.9922	0.9985
11, 19	0.0853	0.1644	0.2830	0.4257	0.5888	0.7286	0.8510	0.9231	0.9711	0.9889	0.9978
11, 20	0.0712	0.1404	0.2500	0.3846	0.5480	0.6908	0.8247	0.9051	0.9631	0.9849	0.9969
12, 12	0.2632	0.4211	0.5789	0.7368	0.8496	0.9301	0.9704	0.9905	0.9972	0.9995	0.9999
12, 13	0.2068	0.3475	0.5000	0.6642	0.7932	0.8937	0.9502	0.9816	0.9939	0.9985	0.9997
12, 14	0.1628	0.2860	0.4296	0.5938	0.7345	0.8518	0.9251	0.9691	0.9886	0.9968	0.9992
12, 15	0.1286	0.2351	0.3681	0.5277	0.6759	0.8062	0.8958	0.9528	0.9813	0.9940	0.9984
12, 16	0.1020	0.1933	0.3149	0.4669	0.6189	0.7585	0.8632	0.9330	0.9718	0.9899	0.9971
12, 17	0.0813	0.1591	0.2693	0.4118	0.5646	0.7101	0.8283	0.9101	0.9602	0.9844	0.9953
12, 18	0.0651	0.1312	0.2304	0.3626	0.5137	0.6621	0.7919	0.8847	0.9465	0.9774	0.9929
12, 19	0.0524	0.1085	0.1973	0.3189	0.4665	0.6153	0.7548	0.8572	0.9311	0.9690	0.9898
12, 20	0.0424	0.0900	0.1693	0.2803	0.4231	0.5703	0.7176	0.8281	0.9140	0.9590	0.9860
13, 13	0.1566	0.2772	0.4179	0.5821	0.7228	0.8434	0.9188	0.9659	0.9869	0.9962	0.9990
13, 14	0.1189	0.2205	0.3475	0.5056	0.6525	0.7880	0.8811	0.9446	0.9764	0.9921	0.9976
13, 15	0.0906	0.1753	0.2883	0.4365	0.5847	0.7299	0.8388	0.9182	0.9623	0.9858	0.9952
13, 16	0.0695	0.1396	0.2389	0.3751	0.5212	0.6714	0.7934	0.8873	0.9446	0.9771	0.9917
13, 17	0.0535	0.1113	0.1980	0.3215	0.4628	0.6141	0.7465	0.8529	0.9238	0.9658	0.9868
13, 18	0.0415	0.0890	0.1643	0.2752	0.4098	0.5592	0.6992	0.8159	0.9001	0.9520	0.9805
13, 19	0.0324	0.0714	0.1365	0.2353	0.3623	0.5074	0.6525	0.7772	0.8742	0.9358	0.9728
13, 20	0.0254	0.0575	0.1138	0.2012	0.3200	0.4592	0.6072	0.7377	0.8465	0.9174	0.9635
14, 14	0.0871	0.1697	0.2798	0.4266	0.5734	0.7202	0.8303	0.9129	0.9588	0.9843	0.9944
14, 15	0.0642	0.1306	0.2247	0.3576	0.5000	0.6519	0.7753	0.8749	0.9358	0.9727	0.9893
14, 16	0.0476	0.1007	0.1804	0.2986	0.4336	0.5854	0.7183	0.8322	0.9081	0.9574	0.9820
14, 17	0.0355	0.0779	0.1450	0.2487	0.3745	0.5226	0.6614	0.7863	0.8765	0.9382	0.9721
14, 18	0.0266	0.0604	0.1167	0.2068	0.3227	0.4643	0.6058	0.7386	0.8418	0.9155	0.9598
14, 19	0.0202	0.0471	0.0942	0.1720	0.2776	0.4110	0.5527	0.6903	0.8049	0.8898	0.9450
14, 20	0.0153	0.0368	0.0763	0.1432	0.2387	0.3630	0.5027	0.6425	0.7667	0.8616	0.9281
15, 15	0.0457	0.0974	0.1749	0.2912	0.4241	0.5759	0.7088	0.8251	0.9026	0.9543	0.9801
15, 16	0.0328	0.0728	0.1362	0.2362	0.3576	0.5046	0.6424	0.7710	0.8638	0.9305	0.9672
15, 17	0.0237	0.0546	0.1061	0.1912	0.3005	0.4393	0.5781	0.7147	0.8210	0.9020	0.9505
15, 18	0.0173	0.0412	0.0830	0.1546	0.2519	0.3806	0.5174	0.6581	0.7754	0.8693	0.9303
15, 19	0.0127	0.0312	0.0650	0.1251	0.2109	0.3286	0.4610	0.6026	0.7285	0.8334	0.9068
15, 20	0.0094	0.0237	0.0512	0.1014	0.1766	0.2831	0.4095	0.5493	0.6813	0.7952	0.8806
16, 16	0.0228	0.0528	0.1028	0.1862	0.2933	0.4311	0.5689	0.7067	0.8138	0.8972	0.9472
16, 17	0.0160	0.0385	0.0778	0.1465	0.2397	0.3659	0.5000	0.6420	0.7603	0.8584	0.9222
16, 18	0.0113	0.0282	0.0591	0.1153	0.1956	0.3091	0.4369	0.5789	0.7051	0.8155	0.8928
16, 19	0.0080	0.0207	0.0450	0.0908	0.1594	0.2603	0.3801	0.5188	0.6498	0.7697	0.8596
16, 20	0.0058	0.0153	0.0345	0.0716	0.1300	0.2188	0.3297	0.4628	0.5959	0.7224	0.8237
17, 17	0.0109	0.0272	0.0572	0.1122	0.1907	0.3028	0.4290	0.5710	0.6972	0.8093	0.8878
17, 18	0.0075	0.0194	0.0422	0.0859	0.1514	0.2495	0.3659	0.5038	0.6341	0.7567	0.8486
17, 19	0.0052	0.0139	0.0313	0.0659	0.1202	0.2049	0.3108	0.4418	0.5728	0.7022	0.8057
17, 20	0.0036	0.0100	0.0233	0.0506	0.0955	0.1680	0.2631	0.3854	0.5146	0.6474	0.7604
18, 18	0.0050	0.0134	0.0303	0.0640	0.1171	0.2004	0.3046	0.4349	0.5651	0.6954	0.7996
18, 19	0.0034	0.0094	0.0219	0.0479	0.0906	0.1606	0.2525	0.3729	0.5000	0.6338	0.7475
18, 20	0.0023	0.0066	0.0159	0.0359	0.0701	0.1285	0.2088	0.3182	0.4398	0.5736	0.6940
19, 19	0.0022	0.0064	0.0154	0.0349	0.0683	0.1256	0.2044	0.3127	0.4331	0.5669	0.6873
19, 20	0.0015	0.0044	0.0109	0.0255	0.0516	0.0981	0.1650	0.2610	0.3729	0.5033	0.6271
20, 20	0.0009	0.0029	0.0075	0.0182	0.0380	0.0748	0.1301	0.2130	0.3143	0.4381	0.5619

12.5. THE RUNS TEST

Distribution of total number of runs v in samples of size (m, n)

m, n	$v = 22$	23	24	25	26	27	28	29
11, 11								
11, 12	1.0000							
11, 13	1.0000							
11, 14	1.0000	1.0000						
11, 15	1.0000	1.0000						
11, 16	0.9999	1.0000						
11, 17	0.9998	1.0000						
11, 18	0.9996	1.0000						
11, 19	0.9994	1.0000						
11, 20	0.9991	1.0000						
12, 12	1.0000							
12, 13	1.0000	1.0000						
12, 14	0.9999	1.0000	1.0000					
12, 15	0.9997	1.0000	1.0000					
12, 16	0.9993	0.9999	1.0000	1.0000				
12, 17	0.9987	0.9998	1.0000	1.0000				
12, 18	0.9978	0.9996	0.9999	1.0000				
12, 19	0.9966	0.9994	0.9999	1.0000				
12, 20	0.9950	0.9991	0.9998	1.0000				
13, 13	0.9998	1.0000	1.0000					
13, 14	0.9995	0.9999	1.0000	1.0000				
13, 15	0.9988	0.9997	1.0000	1.0000				
13, 16	0.9975	0.9994	0.9999	1.0000	1.0000			
13, 17	0.9957	0.9989	0.9997	1.0000	1.0000			
13, 18	0.9930	0.9981	0.9995	0.9999	1.0000	1.0000		
13, 19	0.9894	0.9969	0.9991	0.9999	1.0000	1.0000		
13, 20	0.9848	0.9954	0.9986	0.9998	1.0000	1.0000		
14, 14	0.9985	0.9996	0.9999	1.0000				
14, 15	0.9967	0.9991	0.9998	1.0000	1.0000			
14, 16	0.9938	0.9981	0.9995	0.9999	1.0000	1.0000		
14, 17	0.9894	0.9965	0.9990	0.9998	1.0000	1.0000		
14, 18	0.9834	0.9941	0.9982	0.9996	0.9999	1.0000	1.0000	
14, 19	0.9756	0.9909	0.9970	0.9992	0.9998	1.0000	1.0000	
14, 20	0.9660	0.9867	0.9952	0.9987	0.9997	1.0000	1.0000	
15, 15	0.9930	0.9977	0.9994	0.9999	1.0000	1.0000		
15, 16	0.9872	0.9954	0.9987	0.9997	0.9999	1.0000	1.0000	
15, 17	0.9789	0.9918	0.9974	0.9992	0.9998	1.0000	1.0000	
15, 18	0.9678	0.9866	0.9953	0.9985	0.9996	0.9999	1.0000	1.0000
15, 19	0.9540	0.9798	0.9923	0.9975	0.9993	0.9998	1.0000	1.0000
15, 20	0.9375	0.9712	0.9881	0.9959	0.9987	0.9997	0.9999	1.0000
16, 16	0.9772	0.9908	0.9970	0.9991	0.9998	1.0000	1.0000	
16, 17	0.9634	0.9840	0.9942	0.9981	0.9995	0.9999	1.0000	1.0000
16, 18	0.9457	0.9747	0.9900	0.9964	0.9989	0.9997	0.9999	1.0000
16, 19	0.9244	0.9626	0.9840	0.9938	0.9980	0.9994	0.9999	1.0000
16, 20	0.8996	0.9479	0.9761	0.9903	0.9965	0.9989	0.9997	0.9999
17, 17	0.9428	0.9728	0.9891	0.9959	0.9988	0.9997	0.9999	1.0000
17, 18	0.9172	0.9578	0.9816	0.9925	0.9975	0.9992	0.9998	1.0000
17, 19	0.8872	0.9391	0.9714	0.9876	0.9954	0.9985	0.9996	0.9999
17, 20	0.8534	0.9168	0.9584	0.9808	0.9924	0.9972	0.9992	0.9998
18, 18	0.8829	0.9360	0.9697	0.9866	0.9950	0.9983	0.9995	0.9999
18, 19	0.8438	0.9094	0.9540	0.9781	0.9911	0.9966	0.9990	0.9997
18, 20	0.8010	0.8788	0.9345	0.9670	0.9856	0.9941	0.9980	0.9994
19, 19	0.7956	0.8744	0.9317	0.9651	0.9846	0.9936	0.9978	0.9993
19, 20	0.7444	0.8350	0.9048	0.9484	0.9756	0.9891	0.9959	0.9985
20, 20	0.6857	0.7870	0.8699	0.9252	0.9620	0.9818	0.9925	0.9971

Distribution of total number of runs v in samples of size (m, n)

m, n	$v = 30$	31	32	33	34	35	36	37
16, 16								
16, 17								
16, 18	1.0000							
16, 19	1.0000							
16, 20	1.0000	1.0000						
17, 17	1.0000							
17, 18	1.0000							
17, 19	1.0000	1.0000						
17, 20	0.9999	1.0000						
18, 18	1.0000	1.0000						
18, 19	0.9999	1.0000	1.0000					
18, 20	0.9998	1.0000	1.0000					
19, 19	0.9998	1.0000	1.0000					
19, 20	0.9996	0.9999	1.0000	1.0000				
20, 20	0.9991	0.9997	0.9999	1.0000	1.0000			

The values listed in the previous tables indicate the probability that v or fewer runs will occur. For sample size $m = n$, and m larger than 10, the following table can be used. The columns headed 0.5, 1, 2.5, and 5 give values of v such that v or fewer runs occur with probability less than the indicated percentage. For example, for $m = n = 12$, the probability of 8 or fewer runs is approximately 5%. The columns headed 95, 97.5, 99, and 99.5 give values of v for which the probability of v or more runs is less than 5, 2.5, 1, or 0.5%.

Distribution of the total number of runs v in samples of size $m = n$

$m = n$	0.5	1.0	2.5	5.0	95.0	97.5	99.0	99.5	mean	var (σ^2)	s.d. (σ)
11	5	6	7	7	16	16	17	18	12	5.24	2.29
12	6	7	7	8	17	18	18	19	13	5.74	2.40
13	7	7	8	9	18	19	20	20	14	6.24	2.50
14	7	8	9	10	19	20	21	22	15	6.74	2.60
15	8	9	10	11	20	21	22	23	16	7.24	2.69
16	9	10	11	11	22	22	23	24	17	7.74	2.78
17	10	10	11	12	23	24	25	25	18	8.24	2.87
18	11	11	12	13	24	25	26	26	19	8.74	2.96
19	11	12	13	14	25	26	27	28	20	9.24	3.04
20	12	13	14	15	26	27	28	29	21	9.74	3.12
25	16	17	18	19	32	33	34	35	26	12.24	3.50
30	20	21	23	24	37	38	40	41	31	14.75	3.84
40	29	30	31	33	48	50	51	52	41	19.75	4.44
50	37	38	40	42	59	61	63	64	51	24.75	4.97
60	46	47	49	51	70	72	74	75	61	29.75	5.45
70	55	56	58	60	81	83	85	86	71	34.75	5.89
80	64	65	68	70	91	93	96	97	81	39.75	6.30
90	73	74	77	79	102	104	107	108	91	44.75	6.69
100	82	84	86	88	113	115	117	119	101	49.75	7.05

12.6 THE SIGN TEST

Assumptions: Let $X_1, X_2, \ldots, X_n$ be a random sample from a continuous distribution.

Hypothesis test:

H_0: $\tilde{\mu} = \tilde{\mu}_0$

H_a: $\tilde{\mu} > \tilde{\mu}_0$, $\tilde{\mu} < \tilde{\mu}_0$, $\tilde{\mu} \neq \tilde{\mu}_0$

TS: Y = the number of X_i's greater than $\tilde{\mu}_0$.

Under the null hypothesis, Y has a binomial distribution with parameters n and $p = .5$.

RR: $Y \geq c_1$, $Y \leq c_2$, $Y \geq c$ or $Y \leq n - c$

The critical values c_1, c_2, and c are obtained from the binomial distribution with parameters n and $p = .5$ to yield the desired significance level α. (See the table on page 186.)

Sample values equal to $\tilde{\mu}_0$ are excluded from the analysis and the sample size is reduced accordingly.

The normal approximation: When $n \geq 10$ and $np \geq 5$ the binomial distribution can be approximated by a normal distribution with

$$\mu_Y = np \quad \text{and} \quad \sigma_Y^2 = np(1-p) \tag{12.13}$$

The random variable

$$Z = \frac{Y - \mu_Y}{\sigma_Y} = \frac{Y - np}{\sqrt{np(1-p)}} \tag{12.14}$$

has approximately a standard normal distribution when H_0 is true and $n \geq 10$ and $np \geq 5$.

12.6.1 Table of critical values for the sign test

Let $X_1, X_2, \ldots, X_n$ be a random sample from a continuous distribution with hypothesized median $\tilde{\mu}_0$. The test statistic is Y, the number of X_i's greater than $\tilde{\mu}_0$. If the null hypothesis is true, the probability X_i is greater than the median is $1/2$. The probability distribution for Y is given by the binomial probability function

$$\text{Prob}\,[Y = y] = f(y) = \binom{n}{y}\left(\frac{1}{2}\right)^n. \tag{12.15}$$

The following table contains critical values k such that

$$\text{Prob}\,[Y \leq k] = \sum_{y=0}^{k} \binom{n}{y}\left(\frac{1}{2}\right)^n < \frac{\alpha}{2}. \tag{12.16}$$

For a one-tailed test with significance level α, enter the table in the column headed by 2α.

For larger values of n, approximate critical values may be found using equation (12.14).

$$k = \lfloor np + \sqrt{np(1-p)}z_{\alpha/2} \rfloor \qquad (12.17)$$

where z_α is the critical value for the normal distribution.

Critical values for the sign test

n	probability				n	probability			
	.01	.025	.05	.10		.01	.025	.05	.10
1	0	0	0	0	26	6	6	7	8
2	0	0	0	0	27	6	7	7	8
3	0	0	0	0	28	6	7	8	9
4	0	0	0	0	29	7	8	8	9
5	0	0	0	0	30	7	8	9	10
6	0	0	0	0	31	7	8	9	10
7	0	0	0	0	32	8	9	9	10
8	0	0	0	1	33	8	9	10	11
9	0	0	1	1	34	9	10	10	11
10	0	1	1	1	35	9	10	11	12
11	0	1	1	2	36	9	10	11	12
12	1	1	2	2	37	10	11	12	13
13	1	2	2	3	38	10	11	12	13
14	1	2	2	3	39	11	12	12	13
15	2	2	3	3	40	11	12	13	14
16	2	3	3	4	41	11	12	13	14
17	2	3	4	4	42	12	13	14	15
18	3	3	4	5	43	12	13	14	15
19	3	4	4	5	44	13	14	15	16
20	3	4	5	5	45	13	14	15	16
21	4	4	5	6	46	13	14	15	16
22	4	5	5	6	47	14	15	16	17
23	4	5	6	7	48	14	15	16	17
24	5	6	6	7	49	15	16	17	18
25	5	6	7	7	50	15	16	17	18

12.7 SPEARMAN'S RANK CORRELATION COEFFICIENT

Suppose there are n pairs of observations from continuous distributions. Rank the observations in the two samples separately from smallest to largest. Equal observations are assigned the mean rank for their positions. Let u_i be the rank of the ith observation in the first sample and let v_i be the rank of the ith observation in the second sample. Spearman's rank correlation coefficient, r_S, is a measure of the correlation between ranks, calculated by using the ranks in place of the actual observations in the formula for the correlation coefficient r.

12.7. SPEARMAN'S RANK CORRELATION COEFFICIENT

$$r_S = \frac{SS_{uv}}{\sqrt{SS_{uu} SS_{vv}}} = \frac{n \sum_{i=1}^{n} u_i v_i - \left(\sum_{i=1}^{n} u_i\right)\left(\sum_{i=1}^{n} v_i\right)}{\sqrt{\left[n \sum_{i=1}^{n} u_i^2 - \left(\sum_{i=1}^{n} u_i\right)^2\right]\left[n \sum_{i=1}^{n} v_i^2 - \left(\sum_{i=1}^{n} v_i\right)^2\right]}}$$

$$= 1 - \frac{6 \sum_{i=1}^{n} d_i^2}{n(n^2 - 1)} \quad \text{where } d_i = u_i - v_i. \tag{12.18}$$

The shortcut formula for r_S that only uses the differences $\{d_i\}$ is not exact when there are tied measurements. The approximation is good when the number of ties is small in comparison to n.

Hypothesis test:

H_0: $\rho_S = 0$ (no population correlation between ranks)

H_a: $\rho_S > 0$, $\quad \rho_S < 0$, $\quad \rho_S \neq 0$

TS: r_S

RR: $r_S \geq r_{S,\alpha}$, $\quad r_S \leq -r_{S,\alpha}$, $\quad |r_S| \geq r_{S,\alpha/2}$

where $r_{S,\alpha}$ is a critical value for Spearman's rank correlation coefficient test (see page 188).

The normal approximation: When H_0 is true r_S has approximately a normal distribution with

$$\mu_{r_S} = 0 \quad \text{and} \quad \sigma_{r_S}^2 = \frac{1}{n-1}. \tag{12.19}$$

The random variable

$$Z = \frac{r_S - \mu_{r_S}}{\sigma_{r_S}} = \frac{r_S - 0}{1/\sqrt{n-1}} = r_S \sqrt{n-1} \tag{12.20}$$

has approximately a standard normal distribution as n increases.

12.7.1 Tables for Spearman's rank correlation coefficient

Spearman's coefficient of rank correlation, r_S, measures the correspondence between two rankings; see equation (12.18). The table below gives critical values for r_S assuming the samples are independent; their derivation comes from the subsequent table.

Critical values of Spearman's rank correlation coefficient

n	$\alpha = 0.10$	$\alpha = 0.05$	$\alpha = 0.01$	$\alpha = 0.001$
4	0.8000	0.8000	–	–
5	0.7000	0.8000	0.9000	–
6	0.6000	0.7714	0.8857	–
7	0.5357	0.6786	0.8571	0.9643
8	0.5000	0.6190	0.8095	0.9286
9	0.4667	0.5833	0.7667	0.9000
10	0.4424	0.5515	0.7333	0.8667
11	0.4182	0.5273	0.7000	0.8364
12	0.3986	0.4965	0.6713	0.8182
13	0.3791	0.4780	0.6429	0.7912
14	0.3626	0.4593	0.6220	0.7670
15	0.3500	0.4429	0.6000	0.7464
20	0.2977	0.3789	0.5203	0.6586
25	0.2646	0.3362	0.4654	0.5962
30	0.2400	0.3059	0.4251	0.5479

Let $\sum m$ represents the mean value of the sum of squares. Then the following tables give the probability that $\sum d^2 \geq S$ for $S \geq \sum m$, or that $\sum d^2 \leq S$ for $S \leq \sum m$. The tables for $n = 9$ and $n = 10$ can be completed by symmetry.

The values in the next table create the critical values in the last table. For example, taking $n = 9$ we note that (a) $S = 26$ (corresponding to a Spearman rank correlation coefficient of $1 - \frac{26}{120} \approx 0.7833$) has a probability of 0.0086; and (b) $S = 28$ (corresponding to a Spearman rank correlation coefficient of $1 - \frac{28}{120} \approx 0.7667$) has a probability of 0.0107. Hence, the critical value for $n = 9$ and $\alpha = 0.01$, the least value of the coefficient whose probability is less than 0.01, is 0.7667.

12.7. SPEARMAN'S RANK CORRELATION COEFFICIENT

Exact values for Spearman's rank correlation coefficient

S	$n=2$	3	4	5	6	7	8	9	10
	$\sum m = 1$	4	10	20	35	56	84	120	165
0	0.5000	0.1667	0.0417	0.0083	0.0014	0.0002	0.0000	0.0000	0.0000
2	0.5000	0.5000	0.1667	0.0417	0.0083	0.0014	0.0002	0.0000	0.0000
4		0.5000	0.2083	0.0667	0.0167	0.0034	0.0006	0.0001	0.0000
6		0.5000	0.3750	0.1167	0.0292	0.0062	0.0011	0.0002	0.0000
8		0.1667	0.4583	0.1750	0.0514	0.0119	0.0023	0.0004	0.0001
10			0.5417	0.2250	0.0681	0.0171	0.0036	0.0007	0.0001
12			0.4583	0.2583	0.0875	0.0240	0.0054	0.0010	0.0002
14			0.3750	0.3417	0.1208	0.0331	0.0077	0.0015	0.0003
16			0.2083	0.3917	0.1486	0.0440	0.0109	0.0023	0.0004
18			0.1667	0.4750	0.1778	0.0548	0.0140	0.0030	0.0006
20			0.0417	0.5250	0.2097	0.0694	0.0184	0.0041	0.0008
22				0.4750	0.2486	0.0833	0.0229	0.0054	0.0011
24				0.3917	0.2819	0.1000	0.0288	0.0069	0.0014
26				0.3417	0.3292	0.1179	0.0347	0.0086	0.0019
28				0.2583	0.3569	0.1333	0.0415	0.0107	0.0024
30				0.2250	0.4014	0.1512	0.0481	0.0127	0.0029
32				0.1750	0.4597	0.1768	0.0575	0.0156	0.0036
34				0.1167	0.5000	0.1978	0.0661	0.0184	0.0044
36				0.0667	0.5000	0.2222	0.0756	0.0216	0.0053
38				0.0417	0.5000	0.2488	0.0855	0.0252	0.0063
40				0.0083	0.4597	0.2780	0.0983	0.0294	0.0075
42					0.4014	0.2974	0.1081	0.0333	0.0087
44					0.3569	0.3308	0.1215	0.0380	0.0101
46					0.3292	0.3565	0.1337	0.0429	0.0117
48					0.2819	0.3913	0.1496	0.0484	0.0134
50					0.2486	0.4198	0.1634	0.0540	0.0153
52					0.2097	0.4532	0.1799	0.0603	0.0173
54					0.1778	0.4817	0.1947	0.0664	0.0195
56					0.1486	0.5183	0.2139	0.0738	0.0219
58					0.1208	0.4817	0.2309	0.0809	0.0245
60					0.0875	0.4532	0.2504	0.0888	0.0272
62					0.0681	0.4198	0.2682	0.0969	0.0302
64					0.0514	0.3913	0.2911	0.1063	0.0334
66					0.0292	0.3565	0.3095	0.1149	0.0367
68					0.0167	0.3308	0.3323	0.1250	0.0403
70					0.0083	0.2974	0.3517	0.1348	0.0441
72					0.0014	0.2780	0.3760	0.1456	0.0481
74						0.2488	0.3965	0.1563	0.0524
76						0.2222	0.4201	0.1681	0.0569
78						0.1978	0.4410	0.1793	0.0616
80						0.1768	0.4674	0.1927	0.0667

Exact values for Spearman's rank correlation coefficient

S	$n=2$	3	4	5	6	7	8	9	10
	$\sum m = 1$	4	10	20	35	56	84	120	165
80						0.1768	0.4674	0.1927	0.0667
82						0.1512	0.4884	0.2050	0.0720
84						0.1333	0.5116	0.2183	0.0774
86						0.1179	0.4884	0.2315	0.0831
88						0.1000	0.4674	0.2467	0.0893
90						0.0833	0.4410	0.2603	0.0956
92						0.0694	0.4201	0.2759	0.1022
94						0.0548	0.3965	0.2905	0.1091
96						0.0440	0.3760	0.3067	0.1163
98						0.0331	0.3517	0.3218	0.1237
100						0.0240	0.3323	0.3389	0.1316
102						0.0171	0.3095	0.3540	0.1394
104						0.0119	0.2911	0.3718	0.1478
106						0.0062	0.2682	0.3878	0.1564
108						0.0034	0.2504	0.4050	0.1652
110						0.0014	0.2309	0.4216	0.1744
112						0.0002	0.2139	0.4400	0.1839
114							0.1947	0.4558	0.1935
116							0.1799	0.4742	0.2035
118							0.1634	0.4908	0.2135
120							0.1496	0.5092	0.2241
122							0.1337	0.4908	0.2349
124							0.1215	0.4742	0.2459
126							0.1081	0.4558	0.2567
128							0.0983	0.4400	0.2683
130							0.0855	0.4216	0.2801
132							0.0756	0.4050	0.2918
134							0.0661	0.3878	0.3037
136							0.0575	0.3718	0.3161
138							0.0481	0.3540	0.3284
140							0.0415	0.3389	0.3410
142							0.0347	0.3218	0.3536
144							0.0288	0.3067	0.3665
146							0.0229	0.2905	0.3795
148							0.0184	0.2759	0.3925
150							0.0140	0.2603	0.4056
152							0.0109	0.2467	0.4191
154							0.0077	0.2315	0.4326
156							0.0054	0.2183	0.4458
158							0.0036	0.2050	0.4592
160							0.0023	0.1927	0.4730

12.8 WILCOXON MATCHED-PAIRS SIGNED-RANKS TEST

Assume we have a matched set of n observations $\{x_i, y_i\}$. Let d_i denote the differences $d_i = x_i - y_i$.

Hypothesis test:

H_0: there is no difference in the distribution of the x_i's and the y_i's

H_a: there is a difference

Rank all of the d_i's without regard to sign: the least value of $|d_i|$ gets rank 1, the next largest value gets rank 2, etc. After determining the ranking, affix the signs of the differences to each rank.

TS: $T =$ the smaller sum of the like-signed ranks.

RR: $T \geq c$

where c is found from the table on page 192.

Example 12.57: Suppose $n = 10$ values are as shown in the first two columns of the following table:

x_i	y_i	$d_i = x_i - y_i$	rank of $|d_i|$	signed rank of $|d_i|$
9	8	1	2	2
2	2	0	–	–
1	3	-2	4.5	-4.5
4	2	2	4.5	4.5
6	3	3	7	7
4	0	4	9	9
7	4	3	7	7
8	5	3	7	7
5	4	1	2	2
1	0	1	2	2
				$\sum R^+ = 40.5$
				$\sum R^- = -4.5$

The subsequent columns show the differences, the ranks (note how ties are handled), and the signed ranks. The smaller of the two sums is $T = 4.5$. From the following table (with $n = 10$) we conclude that there is evidence of a difference in distributions at the .005 significance level.

See D. J. Sheskin, *Handbook of Parametric and Nonparametric Statistical Procedures*, CRC Press LLC, Boca Raton, FL, 1997, pages 291–301, 681.

Critical values for the Wilcoxon signed-ranks test and the matched-pairs signed-ranks test

One sided	Two sided	$n=5$	6	7	8	9	10	11	12	13	14
$\alpha=.05$	$\alpha=.10$	0	2	3	5	8	10	13	17	21	25
$\alpha=.025$	$\alpha=.05$		0	2	3	5	8	10	13	17	21
$\alpha=.01$	$\alpha=.02$			0	1	3	5	7	9	12	15
$\alpha=.005$	$\alpha=.01$				0	1	3	5	7	9	12
One sided	Two sided	$n=15$	16	17	18	19	20	21	22	23	24
$\alpha=.05$	$\alpha=.10$	30	35	41	47	53	60	67	75	83	91
$\alpha=.025$	$\alpha=.05$	25	29	34	40	46	52	58	65	73	81
$\alpha=.01$	$\alpha=.02$	19	23	27	32	37	43	49	55	62	69
$\alpha=.005$	$\alpha=.01$	15	19	23	27	32	37	42	48	54	61
One sided	Two sided	$n=25$	26	27	28	29	30	31	32	33	34
$\alpha=.05$	$\alpha=.10$	100	110	119	130	140	151	163	175	187	200
$\alpha=.025$	$\alpha=.05$	89	98	107	116	126	137	147	159	170	182
$\alpha=.01$	$\alpha=.02$	76	84	92	101	110	120	130	140	151	162
$\alpha=.005$	$\alpha=.01$	68	75	83	91	100	109	118	128	138	148

12.9 WILCOXON RANK–SUM (MANN–WHITNEY) TEST

Assumptions: Let $X_1, X_2, \ldots, X_m$ and $Y_1, Y_2, \ldots, Y_n$ (with $m \leq n$) be independent random samples from continuous distributions.

Hypothesis test:

H_0: $\tilde{\mu}_1 - \tilde{\mu}_2 = \Delta_0$

H_a: $\tilde{\mu}_1 - \tilde{\mu}_2 > \Delta_0$, $\quad \tilde{\mu}_1 - \tilde{\mu}_2 < \Delta_0$, $\quad \tilde{\mu}_1 - \tilde{\mu}_2 \neq \Delta_0$

Subtract Δ_0 from each X_i. Combine the $(X_i - \Delta_0)$'s and the Y_j's into one sample and rank all of the observations. Equal differences are assigned the mean rank for their positions.

TS: $W = \sum_{i=1}^{m} R_i$

where R_i is the rank of $(X_i - \Delta_0)$ in the combined sample.

RR: $W \geq c_1$, $\quad W \leq c_2$, $\quad W \geq c$ or $W \leq m(m+n+1) - c$

where c_1, c_2, and c are critical values for the Wilcoxon rank–sum statistic such that $\text{Prob}\,[W \geq c_1] \approx \alpha$, $\text{Prob}\,[W \leq c_2] \approx \alpha$, and $\text{Prob}\,[W \geq c] \approx \alpha/2$. (In practice, we convert from W to U via equation (12.23) and look up U values.)

The normal approximation: When both m and n are greater than 8, W has approximately a normal distribution with

$$\mu_W = \frac{m(m+n+1)}{2} \quad \text{and} \quad \sigma_W^2 = \frac{mn(m+n+1)}{12}. \tag{12.21}$$

12.9. WILCOXON RANK–SUM (MANN–WHITNEY) TEST

The random variable

$$Z = \frac{W - \mu_W}{\sigma_W} \tag{12.22}$$

has approximately a standard normal distribution.

The Mann–Whitney U statistic: The rank–sum test may also be based on the test statistic

$$U = \frac{m(m + 2n + 1)}{2} - W. \tag{12.23}$$

When both m and n are greater than 8, U has approximately a normal distribution with

$$\mu_U = \frac{mn}{2} \quad \text{and} \quad \sigma_U^2 = \frac{mn(m + n + 1)}{12}. \tag{12.24}$$

The random variable

$$Z = \frac{U - \mu_U}{\sigma_U} \tag{12.25}$$

has approximately a standard normal distribution.

Note that there are two tests commonly called the Mann–Whitney U test: one developed by Mann and Whitney and one developed by Wilcoxon. Although they employ different equations and different tables, the two versions yield comparable results.

Example 12.58: The Pennsylvania State Police theorize that cars travel faster during the evening rush hour versus the morning rush hour. Randomly selected cars were selected during each rush hour and their speeds were computed using radar. The data are given in the table below.

Morning:	68	65	80	61	64	64
	63	73	75	71		
Evening:	70	70	71	72	72	71
	75	74	81	72	74	71

Use the Mann–Whitney U test to determine if there is any evidence to suggest the median speeds are different. Use $\alpha = .05$.

Solution:

(S1) Computations:

$m = 10$, $n = 12$, $W = 88$, $U = 87$

(S2) Using the tables, the critical value for a two-sided test with $\alpha = .05$ is 29.

(S3) The value of the test statistic is not in the rejection region. There is no evidence to suggest the median speeds are different.

12.9.1 Tables for Wilcoxon (Mann–Whitney) U statistic

Given two sample of sizes m and n (with $m \leq n$) the Mann–Whitney U-statistic (see equation (12.23)) is used to test the hypothesis that the two samples are from populations with the same median. Rank all of the observations in ascending order of magnitude. Let W be the sum of the ranks assigned to the sample of size m. Then U is defined as

$$U = \frac{m(m+2n+1)}{2} - W \qquad (12.26)$$

The following tables present cumulative probability and are used to determine exact probabilities associated with this test statistic. If the null hypothesis is true, the body of the tables contains probabilities such that $\text{Prob}\,[U \leq u]$.

Only *small* values of u are shown in the tables since the probability distribution for U is symmetric. For example, for $n = 3$ and $m = 2$ the probability distribution of U values is:

$$\text{Prob}\,[U=0] = \text{Prob}\,[U=1] = \text{Prob}\,[U=5] = \text{Prob}\,[U=6] = 0.1$$
$$\text{Prob}\,[U=2] = \text{Prob}\,[U=3] = \text{Prob}\,[U=4] = 0.2$$

so that the distribution function is:

$$\text{Prob}\,[U \leq 0] = 0.1, \quad \text{Prob}\,[U \leq 1] = 0.2, \quad \text{Prob}\,[U \leq 2] = 0.4,$$
$$\text{Prob}\,[U \leq 3] = 0.6, \quad \text{Prob}\,[U \leq 4] = 0.8,$$
$$\text{Prob}\,[U \leq 5] = 0.9, \quad \text{Prob}\,[U \leq 6] = 1$$

Example 12.59: Consider the two samples $\{13, 9\}$ ($m = 2$) and $\{12, 16, 14\}$ ($n = 3$). Arrange the combined samples in rank order and box the values from the first sample:

$$\begin{array}{cccccc} & \boxed{9} & 12 & \boxed{13} & 14 & 16 \\ \text{rank} & 1 & 2 & 3 & 4 & 5 \end{array} \qquad (12.27)$$

Compute the U statistic:

(a) $W = 1 + 3 = 4$

(b) $U = \dfrac{2(2 + 2 \cdot 3 + 1)}{2} - 4 = 5$

Using the tables below (and the comment above): $\text{Prob}\,[U \leq 5] = .9$. There is little evidence to suggest the medians are different.

	$n = 3$		
u	$m = 1$	2	3
0	0.250	0.100	0.050
1	0.500	0.200	0.100
2	0.750	0.400	0.200
3		0.600	0.350
4			0.500
5			0.650

	$n = 4$			
u	$m = 1$	2	3	4
0	0.200	0.067	0.029	0.014
1	0.400	0.133	0.057	0.029
2	0.600	0.267	0.114	0.057
3		0.400	0.200	0.100
4		0.600	0.314	0.171
5			0.429	0.243
6			0.571	0.343
7				0.443
8				0.557

12.9. WILCOXON RANK-SUM (MANN-WHITNEY) TEST

			$n=5$		
u	$m=1$	2	3	4	5
0	0.167	0.048	0.018	0.008	0.004
1	0.333	0.095	0.036	0.016	0.008
2	0.500	0.190	0.071	0.032	0.016
3	0.667	0.286	0.125	0.056	0.028
4		0.429	0.196	0.095	0.048
5		0.571	0.286	0.143	0.075
6			0.393	0.206	0.111
7			0.500	0.278	0.155
8			0.607	0.365	0.210
9				0.452	0.274
10				0.548	0.345
11					0.421
12					0.500
13					0.579

			$n=6$			
u	$m=1$	2	3	4	5	6
0	0.143	0.036	0.012	0.005	0.002	0.001
1	0.286	0.071	0.024	0.010	0.004	0.002
2	0.429	0.143	0.048	0.019	0.009	0.004
3	0.571	0.214	0.083	0.033	0.015	0.008
4		0.321	0.131	0.057	0.026	0.013
5		0.429	0.190	0.086	0.041	0.021
6		0.571	0.274	0.129	0.063	0.032
7			0.357	0.176	0.089	0.047
8			0.452	0.238	0.123	0.066
9			0.548	0.305	0.165	0.090
10				0.381	0.214	0.120
11				0.457	0.268	0.155
12				0.543	0.331	0.197
13					0.396	0.242
14					0.465	0.294
15					0.535	0.350
16						0.409
17						0.469
18						0.531

				$n=7$				
u	$m=1$	2	3	4	5	6	7
0	0.125	0.028	0.008	0.003	0.001	0.001	0.000
1	0.250	0.056	0.017	0.006	0.003	0.001	0.001
2	0.375	0.111	0.033	0.012	0.005	0.002	0.001
3	0.500	0.167	0.058	0.021	0.009	0.004	0.002
4	0.625	0.250	0.092	0.036	0.015	0.007	0.003
5		0.333	0.133	0.055	0.024	0.011	0.006
6		0.444	0.192	0.082	0.037	0.017	0.009
7		0.556	0.258	0.115	0.053	0.026	0.013
8			0.333	0.158	0.074	0.037	0.019
9			0.417	0.206	0.101	0.051	0.027
10			0.500	0.264	0.134	0.069	0.036
11			0.583	0.324	0.172	0.090	0.049
12				0.394	0.216	0.117	0.064
13				0.464	0.265	0.147	0.082
14				0.536	0.319	0.183	0.104
15					0.378	0.223	0.130
16					0.438	0.267	0.159
17					0.500	0.314	0.191
18					0.562	0.365	0.228
19						0.418	0.267
20						0.473	0.310
21						0.527	0.355
22							0.402
23							0.451
24							0.500
25							0.549

12.9.2 Critical values for Wilcoxon (Mann–Whitney) statistic

The following tables give critical values for U for significance levels of 0.00005, 0.0001, 0.005, 0.01, 0.05, and 0.10 for a one-tailed test. For a two-tailed test, the significance levels are doubled. If an observed U is equal to or less than the tabular value, the null hypothesis may be rejected at the level of significance indicated at the head of the table.

12.9. WILCOXON RANK-SUM (MANN-WHITNEY) TEST

Critical values of U in the Mann-Whitney test
Critical values for the $\alpha = 0.10$ level of significance

m	$n=1$	2	3	4	5	6	7	8	9	10	11	12	13	14	15	16	17	18	19	20
1																				
2			0	0	1	1	1	2	2	3	3	4	4	5	5	5	6	6	7	7
3		0	1	1	2	3	4	5	5	6	7	8	9	10	10	11	12	13	14	15
4		0	1	3	4	5	6	7	9	10	11	12	13	15	16	17	18	20	21	22
5		1	2	4	5	7	8	10	12	13	15	17	18	20	22	23	25	27	28	30
6		1	3	5	7	9	11	13	15	17	19	21	23	25	27	29	31	34	36	38
7		1	4	6	8	11	13	16	18	21	23	26	28	31	33	36	38	41	43	46
8		2	5	7	10	13	16	19	22	24	27	30	33	36	39	42	45	48	51	54
9		2	5	9	12	15	18	22	25	28	31	35	38	41	45	48	52	55	58	62
10		3	6	10	13	17	21	24	28	32	36	39	43	47	51	54	58	62	66	70
11		3	7	11	15	19	23	27	31	36	40	44	48	52	57	61	65	69	73	78
12		4	8	12	17	21	26	30	35	39	44	49	53	58	63	67	72	77	81	86
13		4	9	13	18	23	28	33	38	43	48	53	58	63	68	74	79	84	89	94
14		5	10	15	20	25	31	36	41	47	52	58	63	69	74	80	85	91	97	102
15		5	10	16	22	27	33	39	45	51	57	63	68	74	80	86	92	98	104	110
16		5	11	17	23	29	36	42	48	54	61	67	74	80	86	93	99	106	112	119
17		6	12	18	25	31	38	45	52	58	65	72	79	85	92	99	106	113	120	127
18		6	13	20	27	34	41	48	55	62	69	77	84	91	98	106	113	120	128	135
19		7	14	21	28	36	43	51	58	66	73	81	89	97	104	112	120	128	135	143
20		7	15	22	30	38	46	54	62	70	78	86	94	102	110	119	127	135	143	151

Critical values of U in the Mann-Whitney test
Critical values for the $\alpha = 0.05$ level of significance

m	$n=1$	2	3	4	5	6	7	8	9	10	11	12	13	14	15	16	17	18	19	20
1																				
2					0	0	0	1	1	1	1	2	2	3	3	3	3	4	4	4
3			0	0	1	2	2	3	4	4	5	5	6	7	7	8	9	9	10	11
4			0	1	2	3	4	5	6	7	8	9	10	11	12	14	15	16	17	18
5		0	1	2	4	5	6	8	9	11	12	13	15	16	18	19	20	22	23	25
6		0	2	3	5	7	8	10	12	14	16	17	19	21	23	25	26	28	30	32
7		0	2	4	6	8	11	13	15	17	19	21	24	26	28	30	33	35	37	39
8		1	3	5	8	10	13	15	18	20	23	26	28	31	33	36	39	41	44	47
9		1	4	6	9	12	15	18	21	24	27	30	33	36	39	42	45	48	51	54
10		1	4	7	11	14	17	20	24	27	31	34	37	41	44	48	51	55	58	62
11		1	5	8	12	16	19	23	27	31	34	38	42	46	50	54	57	61	65	69
12		2	5	9	13	17	21	26	30	34	38	42	47	51	55	60	64	68	72	77
13		2	6	10	15	19	24	28	33	37	42	47	51	56	61	65	70	75	80	84
14		3	7	11	16	21	26	31	36	41	46	51	56	61	66	71	77	82	87	92
15		3	7	12	18	23	28	33	39	44	50	55	61	66	72	77	83	88	94	100
16		3	8	14	19	25	30	36	42	48	54	60	65	71	77	83	89	95	101	107
17		3	9	15	20	26	33	39	45	51	57	64	70	77	83	89	96	102	109	115
18		4	9	16	22	28	35	41	48	55	61	68	75	82	88	95	102	109	116	123
19		4	10	17	23	30	37	44	51	58	65	72	80	87	94	101	109	116	123	130
20		4	11	18	25	32	39	47	54	62	69	77	84	92	100	107	115	123	130	138

Critical values of U in the Mann–Whitney test
Critical values for the $\alpha = 0.01$ level of significance

m	$n=1$	2	3	4	5	6	7	8	9	10	11	12	13	14	15	16	17	18	19	20
1																				
2										0	0	0	0	0	0	0	0	0	1	1
3						0	0	1	1	1	2	2	2	3	3	4	4	4	4	5
4				0	1	1	2	3	3	4	5	5	6	7	7	8	9	9	10	
5			0	1	2	3	4	5	6	7	8	9	10	11	12	13	14	15	16	
6			1	2	3	4	6	7	8	9	11	12	13	15	16	18	19	20	22	
7		0	1	3	4	6	7	9	11	12	14	16	17	19	21	23	24	26	28	
8		0	2	4	6	7	9	11	13	15	17	20	22	24	26	28	30	32	34	
9		1	3	5	7	9	11	14	16	18	21	23	26	28	31	33	36	38	40	
10		1	3	6	8	11	13	16	19	22	24	27	30	33	36	38	41	44	47	
11		1	4	7	9	12	15	18	22	25	28	31	34	37	41	44	47	50	53	
12		2	5	8	11	14	17	21	24	28	31	35	38	42	46	49	53	56	60	
13	0	2	5	9	12	16	20	23	27	31	35	39	43	47	51	55	59	63	67	
14	0	2	6	10	13	17	22	26	30	34	38	43	47	51	56	60	65	69	73	
15	0	3	7	11	15	19	24	28	33	37	42	47	51	56	61	66	70	75	80	
16	0	3	7	12	16	21	26	31	36	41	46	51	56	61	66	71	76	82	87	
17	0	4	8	13	18	23	28	33	38	44	49	55	60	66	71	77	82	88	93	
18	0	4	9	14	19	24	30	36	41	47	53	59	65	70	76	82	88	94	100	
19	1	4	9	15	20	26	32	38	44	50	56	63	69	75	82	88	94	101	107	
20	1	5	10	16	22	28	34	40	47	53	60	67	73	80	87	93	100	107	114	

Critical values of U in the Mann–Whitney test
Critical values for the $\alpha = 0.005$ level of significance

m	$n=1$	2	3	4	5	6	7	8	9	10	11	12	13	14	15	16	17	18	19	20
1																				
2																			0	0
3						0	0	0	1	1	1	2	2	2	2	3	3			
4				0	0	1	1	2	2	3	3	4	5	5	6	6	7	8		
5			0	1	1	2	3	4	5	6	7	7	8	9	10	11	12	13		
6		0	1	2	3	4	5	6	7	9	10	11	12	13	15	16	17	18		
7		0	1	3	4	6	7	9	10	12	13	15	16	18	19	21	22	24		
8		1	2	4	6	7	9	11	13	15	17	18	20	22	24	26	28	30		
9	0	1	3	5	7	9	11	13	16	18	20	22	24	27	29	31	33	36		
10	0	2	4	6	9	11	13	16	18	21	24	26	29	31	34	37	39	42		
11	0	2	5	7	10	13	16	18	21	24	27	30	33	36	39	42	45	48		
12	1	3	6	9	12	15	18	21	24	27	31	34	37	41	44	47	51	54		
13	1	3	7	10	13	17	20	24	27	31	34	38	42	45	49	53	57	60		
14	1	4	7	11	15	18	22	26	30	34	38	42	46	50	54	58	63	67		
15	2	5	8	12	16	20	24	29	33	37	42	46	51	55	60	64	69	73		
16	2	5	9	13	18	22	27	31	36	41	45	50	55	60	65	70	74	79		
17	2	6	10	15	19	24	29	34	39	44	49	54	60	65	70	75	81	86		
18	2	6	11	16	21	26	31	37	42	47	53	58	64	70	75	81	87	92		
19	0	3	7	12	17	22	28	33	39	45	51	57	63	69	74	81	87	93	99	
20	0	3	8	13	18	24	30	36	42	48	54	60	67	73	79	86	92	99	105	

Critical values of U in the Mann–Whitney test
Critical values for the $\alpha = 0.001$ level of significance

m	n = 1	2	3	4	5	6	7	8	9	10	11	12	13	14	15	16	17	18	19	20
1																				
2																				
3																	0	0	0	0
4								0	0	0	1	1	1	2	2	3	3	3		
5						0	1	1	2	2	3	3	4	5	5	6	7	7		
6					0	1	2	3	4	4	5	6	7	8	9	10	11	12		
7				0	1	2	3	5	6	7	8	9	10	11	13	14	15	16		
8			0	1	2	4	5	6	8	9	11	12	14	15	17	18	20	21		
9			1	2	3	5	7	8	10	12	14	15	17	19	21	23	25	26		
10		0	1	3	5	6	8	10	12	14	17	19	21	23	25	27	29	32		
11		0	2	4	6	8	10	12	15	17	20	22	24	27	29	32	34	37		
12		0	2	4	7	9	12	14	17	20	23	25	28	31	34	37	40	42		
13		1	3	5	8	11	14	17	20	23	26	29	32	35	38	42	45	48		
14		1	3	6	9	12	15	19	22	25	29	32	36	39	43	46	50	54		
15		1	4	7	10	14	17	21	24	28	32	36	40	43	47	51	55	59		
16		2	5	8	11	15	19	23	27	31	35	39	43	48	52	56	60	65		
17	0	2	5	9	13	17	21	25	29	34	38	43	47	52	57	61	66	70		
18	0	3	6	10	14	18	23	27	32	37	42	46	51	56	61	66	71	76		
19	0	3	7	11	15	20	25	29	34	40	45	50	55	60	66	71	77	82		
20	0	3	7	12	16	21	26	32	37	42	48	54	59	65	70	76	82	88		

Critical values of U in the Mann–Whitney test
Critical values for the $\alpha = 0.0005$ level of significance

m	n = 1	2	3	4	5	6	7	8	9	10	11	12	13	14	15	16	17	18	19	20
1																				
2																				
3																				
4														0	0	0	1	1	1	2	2
5							0	0	1	1	2	2	3	3	4	4	5	5		
6						0	1	2	2	3	4	5	5	6	7	8	8	9		
7					0	1	2	3	4	5	6	7	8	9	10	11	13	14		
8				0	1	2	4	5	6	7	9	10	11	13	14	15	17	18		
9			0	1	2	4	5	7	8	10	11	13	15	16	18	20	21	23		
10			0	2	3	5	7	8	10	12	14	16	18	20	22	24	26	28		
11			1	2	4	6	8	10	12	15	17	19	21	24	26	28	31	33		
12			1	3	5	7	10	12	15	17	20	22	25	27	30	33	35	38		
13		0	2	4	6	9	11	14	17	20	23	25	28	31	34	37	40	43		
14		0	2	5	7	10	13	16	19	22	25	29	32	35	39	42	45	49		
15		0	3	5	8	11	15	18	21	25	28	32	36	39	43	46	50	54		
16		1	3	6	9	13	16	20	24	27	31	35	39	43	47	51	55	59		
17		1	4	7	10	14	18	22	26	30	34	39	43	47	51	56	60	65		
18		1	4	8	11	15	20	24	28	33	37	42	46	51	56	61	65	70		
19		2	5	8	13	17	21	26	31	35	40	45	50	55	60	65	70	76		
20		2	5	9	14	18	23	28	33	38	43	49	54	59	65	70	76	81		

12.10 WILCOXON SIGNED-RANK TEST

Assumptions: Let $X_1, X_2, \ldots, X_n$ be a random sample from a continuous symmetric distribution.

Hypothesis test:

H_0: $\tilde{\mu} = \tilde{\mu}_0$

H_a: $\tilde{\mu} > \tilde{\mu}_0$, $\tilde{\mu} < \tilde{\mu}_0$, $\tilde{\mu} \neq \tilde{\mu}_0$

Rank the absolute differences $|X_1 - \tilde{\mu}_0|, |X_2 - \tilde{\mu}_0|, \ldots, |X_n - \tilde{\mu}_0|$. Equal absolute differences are assigned the mean rank for their positions.

TS: T_+ = the sum of the ranks corresponding to the positive differences $(X_i - \tilde{\mu}_0)$.

RR: $T_+ \geq c_1$, $T_+ \leq c_2$, $T_+ \geq c$ or $T_+ \leq n(n+1) - c$

where c_1, c_2, and c are critical values for the Wilcoxon signed-rank statistic (see table on page 192) such that $\text{Prob}\,[T_+ \geq c_1] \approx \alpha$, $\text{Prob}\,[T_+ \leq c_2] \approx \alpha$, and $\text{Prob}\,[T_+ \geq c] \approx \alpha/2$.

Any observed difference $(x_i - \tilde{\mu}_0) = 0$ is excluded from the test and the sample size is reduced accordingly.

The normal approximation: When $n \geq 20$, T_+ has approximately a normal distribution with

$$\mu_{T_+} = \frac{n(n+1)}{4} \quad \text{and} \quad \sigma^2_{T_+} = \frac{n(n+1)(2n+1)}{24}. \tag{12.28}$$

The random variable

$$Z = \frac{T_+ - \mu_{T_+}}{\sigma_{T_+}} \tag{12.29}$$

has approximately a standard normal distribution when H_0 is true.

See D. J. Sheskin, *Handbook of Parametric and Nonparametric Statistical Procedures*, CRC Press LLC, Boca Raton, FL, 1997, pages 83–94.

CHAPTER 13
Miscellaneous topics

The following notation is used throughout this chapter:
(a) $\mathcal{N}$ denotes the *natural numbers*.
(b) $\mathcal{R}$ denotes the *real numbers*.

13.1 CEILING AND FLOOR FUNCTIONS

The ceiling function of x, denoted $\lceil x \rceil$, is the least integer that is not smaller than x. For example, $\lceil \pi \rceil = 4$, $\lceil 5 \rceil = 5$, and $\lceil -1.5 \rceil = -1$.

The floor function of x, denoted $\lfloor x \rfloor$, is the largest integer that is not larger than x. For example, $\lfloor \pi \rfloor = 3$, $\lfloor 5 \rfloor = 5$, and $\lfloor -1.5 \rfloor = -2$.

13.2 ERROR FUNCTIONS

The error function, $\text{erf}(x)$, and the complementary error function, $\text{erfc}(x)$, are defined by:

$$\text{erf}\,x = \frac{2}{\sqrt{\pi}} \int_0^x e^{-t^2}\,dt \qquad \text{erfc}\,x = \frac{2}{\sqrt{\pi}} \int_x^\infty e^{-t^2}\,dt \qquad (13.1)$$

These functions have the properties:
(a) $\text{erf}\,x + \text{erfc}\,x = 1$,
(b) $\text{erf}(-x) = -\text{erf}\,x$,
(c) $\text{erfc}(-x) = 2 - \text{erfc}\,x$

The error function is related to the normal probability distribution function as follows:

$$\Phi(x) = \frac{1}{2}\left[1 + \text{erf}\left(\frac{x}{\sqrt{2}}\right)\right] \qquad (13.2)$$

13.2.1 Special values

Special values of the error function include:

(a) $\text{erf}(\pm\infty) = \pm 1$
(b) $\text{erfc}(-\infty) = 2$
(c) $\text{erfc}(\infty) = 0$
(d) $\text{erf}(0) = 0$
(e) $\text{erfc}(0) = 1$

Note that $\text{erf}(x_0) = \text{erfc}(x_0) = \frac{1}{2}$ for $x_0 = 0.476936\ldots$

13.3 EXPONENTIAL FUNCTION

13.3.1 Exponentiation

For a any real number and m a positive integer, the exponential a^m is defined as

$$a^m = \underbrace{a \cdot a \cdot a \cdots a}_{m \text{ terms}} \tag{13.3}$$

The three laws of exponents are:

1. $a^n \cdot a^m = a^{m+n}$

2. $\dfrac{a^m}{a^n} = \begin{cases} a^{m-n} & \text{if } m > n, \\ 1 & \text{if } m = n, \\ \dfrac{1}{a^{n-m}} & \text{if } m < n. \end{cases}$

3. $(a^m)^n = a^{(mn)}$

The n-th root function is defined as the inverse of the n-th power function. That is

$$\text{if } b^n = a, \text{ then } b = \sqrt[n]{a} = a^{(1/n)}. \tag{13.4}$$

If n is odd, there will be a unique real number satisfying the above definition of $\sqrt[n]{a}$, for any real value of a. If n is even, for positive values of a there will be two real values for $\sqrt[n]{a}$, one positive and one negative. By convention, the symbol $\sqrt[n]{a}$ is understood to mean the positive value. If n is even and a is negative, then there are no real values for $\sqrt[n]{a}$.

To extend the definition to include a^t (for t not necessarily an integer), in such a way as to maintain the laws of exponents, the following definitions are required (where we now restrict a to be positive, p to be an odd number, and q to be an even number):

$$a^0 = 1 \qquad a^{p/q} = \sqrt[q]{a^p} \qquad a^{-t} = \frac{1}{a^t} \tag{13.5}$$

With these restrictions, the second law of exponents can be written as: $\dfrac{a^m}{a^n} = a^{m-n}$.

If $a > 1$ then the function a^x is monotone increasing, while if $0 < a < 1$ then the function a^x is monotone decreasing.

13.3.2 Definition of e^z

$$\exp(z) = e^z = \lim_{m \to \infty} \left(1 + \frac{z}{m}\right)^m = 1 + z + \frac{z^2}{2!} + \frac{z^3}{3!} + \frac{z^4}{4!} + \ldots \tag{13.6}$$

13.4. FACTORIALS AND POCHHAMMER'S SYMBOL

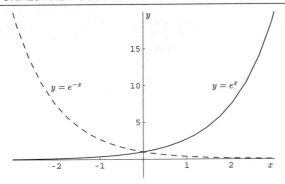

Figure 13.1: Graphs of e^x and e^{-x}.

13.3.3 Derivative and integral of e^z

The derivative of e^z is $\dfrac{de^x}{dx} = e^z$. The integral of e^z is $\displaystyle\int e^x\,dx = e^z + C$.

13.3.4 Circular functions and exponentials

$$\cos z = \frac{e^{iz} + e^{-iz}}{2}, \qquad e^{iz} = \cos z + i\sin z \qquad (13.7)$$

$$\sin z = \frac{e^{iz} - e^{-iz}}{2i} \qquad e^{-iz} = \cos z - i\sin z \qquad (13.8)$$

If $z = x + iy$, then

$$e^z = e^x e^{iy} = e^x(\cos y + i\sin y) \qquad (13.9)$$

13.3.5 Hyperbolic functions

$$\cosh z = \frac{e^z + e^{-z}}{2}, \qquad e^z = \cosh z + \sinh z \qquad (13.10)$$

$$\sinh z = \frac{e^z - e^{-z}}{2} \qquad e^{-z} = \cosh z - \sinh z \qquad (13.11)$$

13.4 FACTORIALS AND POCHHAMMER'S SYMBOL

The factorial of n, denoted $n!$, is the product of all integers less than or equal to n: $n! = n \cdot (n-1) \cdot (n-2) \cdots 2 \cdot 1$. The double factorial of n, denoted $n!!$, is the product of every other integer: $n!! = n \cdot (n-2) \cdot (n-4) \cdots$, where the last element in the product is either 2 or 1, depending on whether n is even or odd. The generalization of the factorial function is the gamma function (see section 13.5). When n is an integer, $\Gamma(n) = (n-1)!$. A table of values is in Table 13.1.

n	$n!$	$\log_{10} n!$	$n!!$	$\log_{10} n!!$
0	1	0.00000	1	0.00000
1	1	0.00000	1	0.00000
2	2	0.30103	2	0.30103
3	6	0.77815	3	0.47712
4	24	1.38021	8	0.90309
5	120	2.07918	15	1.17609
6	720	2.85733	48	1.68124
7	5040	3.70243	105	2.02119
8	40320	4.60552	384	2.58433
9	3.6288×10^5	5.55976	945	2.97543
10	3.6288×10^6	6.55976	3840	3.58433
11	3.9917×10^7	7.60116	10395	4.01682
12	4.7900×10^8	8.68034	46080	4.66351
13	6.2270×10^9	9.79428	1.3514×10^5	5.13077
14	8.7178×10^{10}	10.94041	6.4512×10^5	5.80964
15	1.3077×10^{12}	12.11650	2.0270×10^6	6.30686
16	2.0923×10^{13}	13.32062	1.0322×10^7	7.01376
17	3.5569×10^{14}	14.55107	3.4459×10^7	7.53731
18	6.4024×10^{15}	15.80634	1.8579×10^8	8.26903
19	1.2165×10^{17}	17.08509	6.5473×10^8	8.81606
20	2.4329×10^{18}	18.38612	3.7159×10^9	9.57006
25	1.5511×10^{25}	25.19065	7.9059×10^{12}	12.89795
50	3.0414×10^{64}	64.48307	5.2047×10^{32}	32.71640
100	9.3326×10^{157}	157.97000	3.4243×10^{79}	79.53457
150	5.7134×10^{262}	262.75689	9.3726×10^{131}	131.97186
500	1.2201×10^{1134}	1134.0864	5.8490×10^{567}	567.76709
1000	4.0239×10^{2567}	2567.6046	3.9940×10^{1284}	1284.6014

Table 13.1: Values of the factorial function ($n!$) and the double factorial function ($n!!$).

The shifted factorial (also called the rising factorial or Pochhammer's symbol) is denoted by $(n)_k$ (sometimes $n^{\underline{k}}$), and is defined as

$$(n)_k = \underbrace{n \cdot (n+1) \cdot (n+2) \cdots}_{k \text{ terms}} = \frac{(n+k-1)!}{(n-1)!} = \frac{\Gamma(n+k)}{\Gamma(n)} \qquad (13.12)$$

13.5 GAMMA FUNCTION

The gamma function is defined by

$$\Gamma(z) = \int_0^\infty t^{z-1} e^{-t} \, dt, \quad z = x + iy, \quad x > 0 \qquad (13.13)$$

13.5. GAMMA FUNCTION

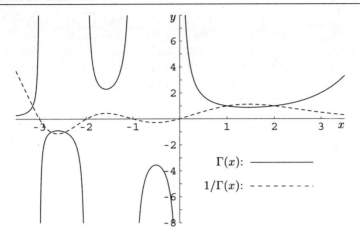

Figure 13.2: Graphs of $\Gamma(x)$ and $1/\Gamma(x)$, for x real.

13.5.1 Properties

$$\Gamma'(1) = \int_0^\infty \ln t \, e^{-t} \, dt = -\gamma \tag{13.14}$$

Multiplication formula:

$$\Gamma(2z) = \pi^{-\frac{1}{2}} 2^{2z-1} \Gamma(z) \Gamma\left(z + \frac{1}{2}\right) \tag{13.15}$$

Reflection formulas:

$$\begin{aligned} \Gamma(z)\Gamma(1-z) &= \frac{\pi}{\sin \pi z}, \\ \Gamma\left(\frac{1}{2}+z\right)\Gamma\left(\frac{1}{2}-z\right) &= \frac{\pi}{\cos \pi z}, \\ \Gamma(z-n) = (-1)^n \Gamma(z) \frac{\Gamma(1-z)}{\Gamma(n+1-z)} &= \frac{(-1)^n \pi}{\sin \pi z \, \Gamma(n+1-z)} \end{aligned} \tag{13.16}$$

The gamma function has the recursion formula:

$$\Gamma(z+1) = z\,\Gamma(z) \tag{13.17}$$

The relation $\Gamma(z) = \Gamma(z+1)/z$ can be used to define the gamma function in the left half plane, $z \neq 0, -1, -2, \ldots$

13.5.2 Expansions

If $z = n$, a large positive integer, then a useful approximation for $n!$ is given by Stirling's formula:

$$n! \sim \sqrt{2\pi n}\, n^n\, e^{-n}, \quad n \to \infty \tag{13.18}$$

n	$\Gamma(n)$	n	$\Gamma(n)$	n	$\Gamma(n)$	n	$\Gamma(n)$
1.00	1.0000	1.25	.9064	1.50	.8862	1.75	.9191
1.05	.9735	1.30	.8975	1.55	.8889	1.80	.9314
1.10	.9514	1.35	.8912	1.60	.8935	1.85	.9456
1.15	.9330	1.40	.8873	1.65	.9001	1.90	.9618
1.20	.9182	1.45	.8857	1.70	.9086	1.95	.9799
1.25	.9064	1.50	.8862	1.75	.9191	2.00	1.0000

Table 13.2: Values of the gamma function.

13.5.3 Special values

$$\Gamma(n+1) = n! \quad \text{if } n = 0, 1, 2, \ldots, \text{where } 0! = 1,$$

$$\Gamma(1) = 1, \quad \Gamma(2) = 1, \quad \Gamma(3) = 2, \quad \Gamma\left(\frac{1}{2}\right) = \sqrt{\pi},$$

$$\Gamma\left(m + \frac{1}{2}\right) = \frac{1 \cdot 3 \cdot 5 \cdots (2m-1)}{2^m} \sqrt{\pi}, \quad m = 1, 2, 3, \ldots,$$

$$\Gamma\left(-m + \frac{1}{2}\right) = \frac{(-1)^m 2^m}{1 \cdot 3 \cdot 5 \cdots (2m-1)} \sqrt{\pi}, \quad m = 1, 2, 3, \ldots$$

$\Gamma(\frac{1}{4}) = 3.62560\,99082$ $\qquad$ $\Gamma(\frac{1}{3}) = 2.67893\,85347$

$\Gamma(\frac{1}{2}) = \sqrt{\pi} = 1.77245\,38509$ $\qquad$ $\Gamma(\frac{2}{3}) = 1.35411\,79394$

$\Gamma(\frac{3}{4}) = 1.22541\,67024$ $\qquad$ $\Gamma(\frac{3}{2}) = \sqrt{\pi}/2 = 0.88622\,69254$

See table 13.2.

13.6 HYPERGEOMETRIC FUNCTIONS

The generalized hypergeometric function is defined by:

$$\begin{aligned}
{}_pF_q\left(a_1, a_2, \ldots, a_p; b_1, b_2, \ldots, b_q; x\right) &= {}_pF_q\left[\begin{array}{c} a_1, \ldots, a_p; x \\ b_1, \ldots, b_q \end{array}\right] \\
&= \sum_{k=0}^{\infty} \frac{\prod_{i=1}^{p}(a_i)_k}{\prod_{i=1}^{q}(b_i)_k} \frac{x^k}{k!} \quad (13.19) \\
&= \sum_{k=0}^{\infty} \frac{(a_1)_k \cdots (a_p)_k}{(b_1)_k \cdots (b_q)_k} \frac{x^k}{k!}
\end{aligned}$$

where $(n)_k$ represents Pochhammer's symbol. Usually, ${}_2F_1(a, b; c; x)$ is called "the" hypergeometric function; this is also called the Gauss hypergeometric function.

13.7 LOGARITHMIC FUNCTIONS

13.7.1 Definition of the natural log

The natural logarithm (also known as the Naperian logarithm) of z is written as $\ln z$ or as $\log_e z$. It is sometimes written $\log z$ (this is also used to represent a "generic" logarithm, a logarithm to any base). One definition is

$$\ln z = \int_1^z \frac{dt}{t}, \qquad (13.20)$$

where the integration path from 1 to z does not cross the origin or the negative real axis.

For complex values of z the natural logarithm, as defined above, can be represented in terms of its magnitude and phase. If $z = x + iy = re^{i\theta}$, then $\ln z = \ln r + i\theta$, where $r = \sqrt{x^2 + y^2}$, $x = r\cos\theta$, and $y = r\sin\theta$.

13.7.2 Special values

$$\lim_{\epsilon \to 0}(\ln \epsilon) = -\infty \qquad \ln 1 = 0 \qquad \ln e = 1$$

$$\ln(-1) = i\pi + 2\pi i k \qquad \ln(\pm i) = \pm\frac{i\pi}{2} + 2\pi i k$$

13.7.3 Logarithms to a base other than e

The logarithmic function to the base a, written $\log_a$, is defined as

$$\log_a z = \frac{\log_b z}{\log_b a} = \frac{\ln z}{\ln a} \qquad (13.21)$$

Note the properties:

(a) $\log_a a^p = p$

(b) $\log_a b = \dfrac{1}{\log_b a}$

(c) $\log_{10} z = \dfrac{\ln z}{\ln 10} = (\log_{10} e)\ln z \approx (0.4342944819\ldots)\ln z$

(d) $\ln z = (\ln 10)\log_{10} z \approx (2.3025850929\ldots)\log_{10} z$

13.7.4 Relation of the logarithm to the exponential

For real values of z the *logarithm* is a monotonic function, as is the exponential. Any monotonic function has a single-valued inverse function; the natural logarithm is the inverse of the exponential. That is, if $x = e^y$, then $y = \ln x$ and $x = e^{\ln x}$. The same inverse relations exist for bases other than e. For example, if $u = a^w$, then $w = \log_a u$ and $u = a^{\log_a u}$.

13.7.5 Identities

$$\log_a z_1 z_2 = \log_a z_1 + \log_a z_2 \qquad \text{for } (-\pi < \arg z_1 + \arg z_2 < \pi)$$
$$\log_a \frac{z_1}{z_2} = \log_a z_1 - \log_a z_2 \qquad \text{for } (-\pi < \arg z_1 - \arg z_2 < \pi)$$
$$\log_a z^n = n \log_a z \qquad \text{for } (-\pi < n \arg z < \pi), \text{ when } n \text{ is an integer}$$

13.7.6 Series expansions for the natural logarithm

$$\ln(1+z) = z - \frac{1}{2}z^2 + \frac{1}{3}z^3 - \cdots, \qquad \text{for } |z| < 1$$

$$\ln z = \left(\frac{z-1}{z}\right) + \frac{1}{2}\left(\frac{z-1}{z}\right)^2 + \frac{1}{3}\left(\frac{z-1}{z}\right)^3 + \cdots \quad \text{for } \operatorname{Re}(z) \geq \frac{1}{2}$$

13.7.7 Derivative and integration formulae

$$\frac{d \ln z}{dz} = \frac{1}{z} \qquad \int \frac{dz}{z} = \ln|z| + C \qquad \int \ln z \, dz = z \ln|z| - z + C \quad (13.22)$$

13.8 SUMS OF POWERS OF INTEGERS

Define $s_k(n) = 1^k + 2^k + \cdots + n^k = \sum_{m=1}^{n} m^k$. Properties include:

(a) $s_k(n) = (k+1)^{-1}[B_{k+1}(n+1) - B_{k+1}(0)]$, where the B_k are Bernoulli polynomials

(b) Writing $s_k(n)$ as $\sum_{m=1}^{k+1} a_m n^{k-m+2}$ there is the recursion formula:

$$s_{k+1}(n) = \binom{k+1}{k+2} a_1 n^{k+2} + \cdots + \binom{k+1}{k} a_3 n^k$$
$$+ \cdots + \binom{k+1}{2} a_{k+1} n^2 + \left[1 - (k+1) \sum_{m=1}^{k+1} \frac{a_m}{k+3-m}\right] n \quad (13.23)$$

$$s_1(n) = 1 + 2 + 3 + \cdots + n = \frac{1}{2}n(n+1)$$
$$s_2(n) = 1^2 + 2^2 + 3^2 + \cdots + n^2 = \frac{1}{6}n(n+1)(2n+1)$$
$$s_3(n) = 1^3 + 2^3 + 3^3 + \cdots + n^3 = \frac{1}{4}(n^2(n+1)^2) = [s_1(n)]^2$$
$$s_4(n) = 1^4 + 2^4 + 3^4 + \cdots + n^4 = \frac{1}{5}(3n^2 + 3n - 1)s_2(n)$$

13.8. SUMS OF POWERS OF INTEGERS

n	$\sum_{k=1}^{n} k$	$\sum_{k=1}^{n} k^2$	$\sum_{k=1}^{n} k^3$	$\sum_{k=1}^{n} k^4$	$\sum_{k=1}^{n} k^5$
1	1	1	1	1	1
2	3	5	9	17	33
3	6	14	36	98	276
4	10	30	100	354	1300
5	15	55	225	979	4425
6	21	91	441	2275	12201
7	28	140	784	4676	29008
8	36	204	1296	8772	61776
9	45	285	2025	15333	120825
10	55	385	3025	25333	220825
11	66	506	4356	39974	381876
12	78	650	6084	60710	630708
13	91	819	8281	89271	1002001
14	105	1015	11025	127687	1539825
15	120	1240	14400	178312	2299200
16	136	1496	18496	243848	3347776
17	153	1785	23409	327369	4767633
18	171	2109	29241	432345	6657201
19	190	2470	36100	562666	9133300
20	210	2870	44100	722666	12333300

Table 13.3: Sums of powers of integers.

$$s_5(n) = 1^5 + 2^5 + 3^5 + \cdots + n^5 = \frac{1}{12}n^2(n+1)^2(2n^2+2n-1)$$

$$s_6(n) = \frac{n}{42}(n+1)(2n+1)(3n^4+6n^3-3n+1)$$

$$s_7(n) = \frac{n^2}{24}(n+1)^2(3n^4+6n^3-n^2-4n+2)$$

$$s_8(n) = \frac{n}{90}(n+1)(2n+1)(5n^6+15n^5+5n^4-15n^3-n^2+9n-3)$$

$$s_9(n) = \frac{n^2}{20}(n+1)^2(2n^6+6n^5+n^4-8n^3+n^2+6n-3)$$

$$s_{10}(n) = \frac{n}{66}(n+1)(2n+1)(3n^8+12n^7+8n^6-18n^5 \\ -10n^4+24n^3+2n^2-15n+5)$$

13.8.1 Permutations

This table contains the number of permutations of n distinct things taken m at a time, given by (see section 3.2.3):

$$P(n,m) = \frac{n!}{(n-m)!} = n(n-1)\cdots(n-m+1) \qquad (13.24)$$

Permutations $P(n,m)$

n	$m=0$	1	2	3	4	5	6	7	8
0	1								
1	1	1							
2	1	2	2						
3	1	3	6	6					
4	1	4	12	24	24				
5	1	5	20	60	120	120			
6	1	6	30	120	360	720	720		
7	1	7	42	210	840	2520	5040	5040	
8	1	8	56	336	1680	6720	20160	40320	40320
9	1	9	72	504	3024	15120	60480	181440	362880
10	1	10	90	720	5040	30240	151200	604800	1814400
11	1	11	110	990	7920	55440	332640	1663200	6652800
12	1	12	132	1320	11880	95040	665280	3991680	19958400
13	1	13	156	1716	17160	154440	1235520	8648640	51891840
14	1	14	182	2184	24024	240240	2162160	17297280	121080960
15	1	15	210	2730	32760	360360	3603600	32432400	259459200

Permutations $P(n,m)$

n	$m=9$	10	11	12	13
9	362880				
10	3628800	3628800			
11	19958400	39916800	39916800		
12	79833600	239500800	479001600	479001600	
13	259459200	1037836800	3113510400	6227020800	6227020800
14	726485760	3632428800	14529715200	43589145600	87178291200
15	1816214400	10897286400	54486432000	217945728000	653837184000

13.8.2 Combinations

This table contains the number of combinations of n distinct things taken m at a time, given by (see section 3.2.5):

$$C(n,m) = \binom{n}{m} = \frac{n!}{m!(n-m)!} \qquad (13.25)$$

13.8. SUMS OF POWERS OF INTEGERS

Combinations $C(n,m)$

n	$m=0$	1	2	3	4	5	6	7
1	1	1						
2	1	2	1					
3	1	3	3	1				
4	1	4	6	4	1			
5	1	5	10	10	5	1		
6	1	6	15	20	15	6	1	
7	1	7	21	35	35	21	7	1
8	1	8	28	56	70	56	28	8
9	1	9	36	84	126	126	84	36
10	1	10	45	120	210	252	210	120
11	1	11	55	165	330	462	462	330
12	1	12	66	220	495	792	924	792
13	1	13	78	286	715	1287	1716	1716
14	1	14	91	364	1001	2002	3003	3432
15	1	15	105	455	1365	3003	5005	6435
16	1	16	120	560	1820	4368	8008	11440
17	1	17	136	680	2380	6188	12376	19448
18	1	18	153	816	3060	8568	18564	31824
19	1	19	171	969	3876	11628	27132	50388
20	1	20	190	1140	4845	15504	38760	77520
21	1	21	210	1330	5985	20349	54264	116280
22	1	22	231	1540	7315	26334	74613	170544
23	1	23	253	1771	8855	33649	100947	245157
24	1	24	276	2024	10626	42504	134596	346104
25	1	25	300	2300	12650	53130	177100	480700
26	1	26	325	2600	14950	65780	230230	657800
27	1	27	351	2925	17550	80730	296010	888030
28	1	28	378	3276	20475	98280	376740	1184040
29	1	29	406	3654	23751	118755	475020	1560780
30	1	30	435	4060	27405	142506	593775	2035800
31	1	31	465	4495	31465	169911	736281	2629575
32	1	32	496	4960	35960	201376	906192	3365856
33	1	33	528	5456	40920	237336	1107568	4272048
34	1	34	561	5984	46376	278256	1344904	5379616
35	1	35	595	6545	52360	324632	1623160	6724520
36	1	36	630	7140	58905	376992	1947792	8347680
37	1	37	666	7770	66045	435897	2324784	10295472
38	1	38	703	8436	73815	501942	2760681	12620256
39	1	39	741	9139	82251	575757	3262623	15380937
40	1	40	780	9880	91390	658008	3838380	18643560
41	1	41	820	10660	101270	749398	4496388	22481940
42	1	42	861	11480	111930	850668	5245786	26978328
43	1	43	903	12341	123410	962598	6096454	32224114
44	1	44	946	13244	135751	1086008	7059052	38320568
45	1	45	990	14190	148995	1221759	8145060	45379620
46	1	46	1035	15180	163185	1370754	9366819	53524680
47	1	47	1081	16215	178365	1533939	10737573	62891499
48	1	48	1128	17296	194580	1712304	12271512	73629072
49	1	49	1176	18424	211876	1906884	13983816	85900584
50	1	50	1225	19600	230300	2118760	15890700	99884400

Combinations $C(n, m)$

n	$m = 8$	9	10	11	12
8	1				
9	9	1			
10	45	10	1		
11	165	55	11	1	
12	495	220	66	12	1
13	1287	715	286	78	13
14	3003	2002	1001	364	91
15	6435	5005	3003	1365	455
16	12870	11440	8008	4368	1820
17	24310	24310	19448	12376	6188
18	43758	48620	43758	31824	18564
19	75582	92378	92378	75582	50388
20	125970	167960	184756	167960	125970
21	203490	293930	352716	352716	293930
22	319770	497420	646646	705432	646646
23	490314	817190	1144066	1352078	1352078
24	735471	1307504	1961256	2496144	2704156
25	1081575	2042975	3268760	4457400	5200300
26	1562275	3124550	5311735	7726160	9657700
27	2220075	4686825	8436285	13037895	17383860
28	3108105	6906900	13123110	21474180	30421755
29	4292145	10015005	20030010	34597290	51895935
30	5852925	14307150	30045015	54627300	86493225
31	7888725	20160075	44352165	84672315	141120525
32	10518300	28048800	64512240	129024480	225792840
33	13884156	38567100	92561040	193536720	354817320
34	18156204	52451256	131128140	286097760	548354040
35	23535820	70607460	183579396	417225900	834451800
36	30260340	94143280	254186856	600805296	1251677700
37	38608020	124403620	348330136	854992152	1852482996
38	48903492	163011640	472733756	1203322288	2707475148
39	61523748	211915132	635745396	1676056044	3910797436
40	76904685	273438880	847660528	2311801440	5586853480
41	95548245	350343565	1121099408	3159461968	7898654920
42	118030185	445891810	1471442973	4280561376	11058116888
43	145008513	563921995	1917334783	5752004349	15338678264
44	177232627	708930508	2481256778	7669339132	21090682613
45	215553195	886163135	3190187286	10150595910	28760021745
46	260932815	1101716330	4076350421	13340783196	38910617655
47	314457495	1362649145	5178066751	17417133617	52251400851
48	377348994	1677106640	6540715896	22595200368	69668534468
49	450978066	2054455634	8217822536	29135916264	92263734836
50	536878650	2505433700	10272278170	37353738800	121399651100

List of Notation

Symbols

!!: double factorial 203
!: factorial 203
$(n)_k$: Pochhammer's symbol 204
$\binom{n}{k}$: binomial coefficient 20
$\binom{n}{n_1,\ldots,n_k}$: multinomial coefficient 22
$\lceil\ \rceil$: ceiling function 201
$\lfloor\ \rfloor$: floor function 201
$\bar{\ }$: mean 7
|: conditional probability 25
$\sim$: distribution similarity 164
$\tilde{\ }$: median 9

Greek Letters

α: Weibull parameter 109
α: confidence coefficient 141
α: type I error 147
β: Weibull parameter 109
β: type II error 147
ϵ: error of estimation 135
$\hat{e}_{ij}$: estimated expected count 153
$\Gamma(x)$: gamma function 204
κ_r: cumulant 32
λ: parameter
    exponential distribution 88
    Poisson distribution 72
λ: test statistic 151
μ: parameter
    location 99
    scale 97
μ_r: moment about the mean 30
$\mu_{[r]}$: factorial moment 30
μ'_r: moment about the origin 30
ν: parameter
    t distribution 104
    chi distribution 81
    chi-squared distribution 80
ν_1: parameter
    F distribution 90
ν_2: parameter
    F distribution 90
ϕ: characteristic function 32
$\Phi(z)$: normal distribution function 116
ρ_{ij}: correlation coefficient 36
σ: parameter
    Rayleigh distribution 103
    scale 99
    shape 97
σ: standard deviation 29
σ^2: variance 29
σ_i: standard deviation 36
σ_{ii}: variance 36
σ_{ij}: covariance 36
τ: Kendall's Tau 166
θ: distribution parameter 135
θ: shape parameter 102

A

A: midrange 50
a: location parameter 102

B

$B[\ ]$: bias 135

C

c: bin width 7
$C(n,k)$: k-combination 21
$C(n;n_1,\ldots,n_k)$: multinomial
    coefficient 22
cdf: cumulative distribution function 27
CF: cumulative frequency 10
$\cosh(x)$: hyperbolic function 203
$\cos(x)$: circular function 203
CQV: coefficient of quartile variation 15

213

$C^R(n,k)$: k-combination with replacement 21
$c(t)$: cumulant generating function 32
CV: coefficient of variation 15

D

D: Kolmogorov–Smirnoff statistic 168, 169
D^+: Kolmogorov–Smirnoff statistic 168, 169
D_n: derangement 23

E

$E[\,]$: expectation 29
e_i: residual 158
erf: error function 201
erfc: complementary error function 201

F

f_k: frequency 7
$F(x)$: cumulative distribution function 27
$F(x_1, x_2, \ldots, x_n)$: cumulative distribution function 33
$f(x_1, x_2, \ldots, x_n)$: probability density function 33

G

g_1: coefficient of skewness 16
g_2: coefficient of skewness 17
GM: geometric mean 8

H

H_0: null hypothesis 147
H_a: alternative hypothesis 147
HM: harmonic mean 8

I

iid: independent and identically distributed 44
IQR: interquartile range 13

J

J: determinant of the Jacobian ... 42

L

$L(\theta)$: likelihood function 151

ln: logarithm 207
log: logarithm 207
$\log_b$: logarithm to base b 207

M

MD: mean deviation 11
mgf: moment generating function 30
MLE: maximum likelihood estimator 137
M_o: mode 9
m_r: moment about the mean 15
m_r': moment about the origin 15
MSE: mean square—error .. 135, 161
MSR: mean square—regression .. 161
MVUE: minimum variance unbiased estimator 136
$m_X(t)$: moment generating function 30

N

$\mathcal{N}$: natural numbers 201
n: shape parameter 87

P

$P(n,k)$: k-permutation 21
p-value 148
pdf: probability density function .. 28
pmf: probability mass function ... 27
$P^R(n,k)$: k-permutation with replacement 21
Prob[ ]: probability 24
$P(t)$: factorial moment generating function 31
$p(x)$: probability mass function ... 27

Q

QD: quartile deviation 13
Q_i: i^{th} quartile 10

R

R: range 50
$\mathcal{R}$: real numbers 201
r: sample correlation coefficient 162
RMS: root mean square 13
RR: rejection region 147
r_S: Spearman's rank correlation coefficient 187

LIST OF NOTATION

$r_{S,\alpha}$: Spearman's rank correlation coefficient 187

S

S: sample space 23
s: sample standard deviation 12
S^2: sample variance 45
s^2 sample variance 11
SEM: standard error—mean 12
$\sinh(x)$: hyperbolic function 203
$\sin(x)$: circular function 203
S_{k_1}: coefficient of skewness 16
S_{k_2}: coefficient of skewness 16
S_{k_q}: coefficient of skewness 16
SSE: sum of squares—error 159
SSLF: sum of squares—lack of fit 162
SSPE: sum of squares—pure error 162
SSR: sum of squares—regression 159
SST: sum of squares—total 159
S_t: Kendall's score 165
S_{xx} 158
S_{xy} 158
S_{yy} 158

T

T: sample total 45
TS: test statistic 147

U

U: Mann–Whitney U statistic ... 193
u_i: coded class mark 7
UMV: uniformly minimum variance unbiased 139

V

VR: variance ratio 173

W

W: range, standardized 53
W: range, studentized 54
w_i: weight 8

X

$X_{(i)}$: order statistic 48
$\overline{X}$: sample mean 45
$\bar{x}$: mean of x 7
x_i: class mark 7
x_o: computing origin 7
$\tilde{x}$: median of x 9
$\overline{x}_{\mathrm{tr}(p)}$: trimmed mean of x 10

Index

A

algebra of sets 19
alternative hypothesis 147
anova 173
    table 161
approximation to
    deciles 11
    median 9
    mode 9
    percentiles 11
    quartiles 11
arcsin distribution 78
arithmetic mean 7, 8
arrangements 23
axioms of probability 24

B

balls into cells 21
bar chart 6
Bayes theorem 26
Bernoulli
    distribution 61
    definition 59
    numbers 205
    polynomials 208
    trials 59, 60, 66, 70
Bessel functions 104
beta
    distribution 50, 97
    function 77
bias 135
biased coin 175
Bienaymé–Chebyshev inequality .. 38
binomial
    coefficients 20
    distribution 60, 69, 70, 142, 146, 185
        definition 60
        tables 61

bivariate distribution 131, 132
    normal 162
box plots 13

C

Cauchy
    –Schwartz inequality 38
    distribution ... 41, 100, 105, 109
        definition 79
ceiling function 201
cells and balls 21
central limit theorem 45
central tendency 9
characteristic
    curves 128
    function 32, 43, 44, 77, 131
chart
    bar 6
    pie 6
Chebyshev
    inequality 38, 46
    one-sided 38
chi distribution 81, 104
chi-square
    distribution .. 41, 47, 48, 88, 90, 97, 100, 102, 110, 112, 151–153, 173
        definition 80
    theorems 47
circular
    functions 203
    permutations 20
coded data 7
coefficient
    binomial 20
    correlation 36, 162
        Kendall's rank 165, 166
        Spearman's rank ... 186, 187
    multinomial 22

coefficient of
- determination 159
- excess kurtosis 17
- kurtosis 17
- momental skewness 16
- quartile variation 15
- skewness 16
- variation 15

coin
- biased 61, 67, 71, 175
- fair 25
- flipping 175

column total 153
combinations 20, 21, 210
combinatorial methods 19
common critical values 142
complementary error function ... 201
complete statistic 136
compound event 23
computing origin 7
conditional
- distribution 35
- probability 25

confidence
- coefficient 141
- interval 141, 159, 160
  - median 144
  - median, difference 145
  - one-sided 141
  - summary 143

consistent estimator 135, 137
contingency table 153, 154
continuous random variables 28
correction
- factor 46
- finite population 146

correlation
- coefficient 36, 165, 166, 186, 187
- sample 162

covariance 36, 46
Cramér-Rao inequality 136
critical values 141, 142, 154–156
- defined 147
- Kolmogorov-Smirnoff test 170, 171
- Mann–Whitney test 196
- normal RV 125
- sign test 185

Wilcoxon test 193, 196
cumulant 32, 131
- generating function 32

cumulative
- distribution function 27, 28
- frequency polygon 6

D

data
- sets 1
- soda pop 1
- summarizing 3
- swimming pool 1
- ticket 1
- transformations 17

deciles 11
definitions 135
degrees of freedom 80–82, 90, 91, 97, 100, 105, 110, 112, 113
density function 28
derangements 23
design matrix 164

deviation
- mean 11
- standard 12

dice 25, 75

discrete
- random variables 27
- uniform distribution 75

distribution
- arcsin 78
- Bernoulli 59, 61
- beta 50, 97
- binomial ... 60, 69, 70, 142, 146, 185
- bivariate 131, 132
- Cauchy 41, 79, 100, 105, 109
- chi 81, 104
- chi-square 41, 47, 48, 80, 88, 90, 97, 100, 102, 110, 112, 151–153, 173
- conditional 35
- continuous 77
- discrete 59
- Erlang 87, 89, 97
- exponential ... 67, 72, 80, 87–89, 96, 102, 104, 108, 109, 113
- extreme-value 89, 109
- F .. 47, 80, 89, 90, 105, 113, 173

INDEX

distribution (*continued*)
    -free statistics 165
    function 27, 28
    gamma 96, 97
    geometric 66, 67, 71
    hypergeometric 68
    joint 49
    Laplace 89
    logistic 89, 102
    lognormal 97, 100
    marginal 34
    means 45
    midrange 50
    multinomial 70
    multivariate 32, 101, 131
    negative binomial 67, 70, 71
    noncentral
        chi-square 100
        t 113
    normal 40–42, 45, 47, 48, 51,
        54, 61, 72, 79, 97, 99, 101,
        105, 110, 112, 113, 115,
        131, 132, 139, 145, 146,
        152, 157, 158, 160, 161,
        164, 175, 176, 185, 187,
        192, 193, 200, 201
        bivariate 162
    Pareto 89, 102
    Pascal 67
    Poisson .. 61, 69, 71, 72, 82, 139
    power function 78, 89, 102
    range 50, 53
        studentized 53
    ratio 41
    Rayleigh ... 43, 81, 88, 100, 103,
        109
    relationships 110
    sampling 44
    studentized range 53
    t 47, 81, 91, 104, 110, 112,
        133, 159–161
        noncentral 113
    triangular 107, 109
    uniform 41, 43, 50, 78, 89,
        108, 113, 139
    waiting time 66
    Weibull 88, 109
double factorial 203

E

efficiency 135, 165
equally likely 24
equations
    between means 9
    normal 157, 164
erf function 103
Erlang
    distribution 89, 97
    definition 87
error
    function 103, 201
    of estimation 135, 142
    probable 12
    type I 147
    type II 147
estimation 135
estimator
    consistent 135, 137
    error 135
    least squares 158, 164
    maximum likelihood 138
    minimum variance 139
    sufficient 135
    unbiased 135, 136
event 23
    compound 23
    rare 72
    simple 23
examples .. 3–5, 10, 16, 17, 20–23, 25,
    26, 34, 35, 39–42, 44, 51,
    61, 67, 69, 71, 72, 75, 82,
    91, 105, 126, 136, 138, 142,
    143, 148, 150, 152–154,
    162, 166, 168, 174, 175,
    191, 193, 194
expansions 205
expectation 29, 33
exponential
    distribution .. 67, 72, 80, 87, 89,
    96, 102, 104, 108, 109, 113
    definition 88
    function 202, 207
exponentiation 202
extreme-value distribution ... 89, 109

F

F distribution .. 80, 90, 105, 113, 173
    definition 89
    theorems 47
factor
    correction 46
    tolerance 125
factorial
    double 203
    moments 30, 31, 44
    relation to gamma function 206
    rising 204
    table 204
failure 59, 68
    number 70
fences 13
finite population 46
finite population correction 146
first order statistic 49
floor function 201
fourth spread 13
frequency distribution 3
frequency polygons 5
Friedman test 165
functions of a random variable
    max 48
    midrange 50
    min 48, 67
    range 50

G

gamma
    distribution 96, 97
        definition 96
    function 77, 80, 81, 90, 96, 104, 109, 132
        definition 204
generating functions ... 30–32, 36, 43, 44
geometric
    distribution 67, 71
        definitions 66
        tables 67
    mean 8
goodness of fit 167
    test 151
graphical procedures 3
grouped data 7

grouping correction 17

H

harmonic mean 8
histogram 4
    probability 5
hyperbolic functions 203
hypergeometric distribution
    definition 68
hypergeometric function .. 69, 77, 90, 102, 132, 206
hypothesis
    alternative 147
    null 147
    test 147, 159–162, 165, 173, 175, 185, 187, 191, 192, 200

I

independence 27, 153
independent random variables 29, 34, 36
inequalities
    Bienaymé–Chebyshev 38
    Cauchy–Schwartz 38
    Chebyshev 38, 46
        one-sided 38
    Cramér–Rao 136
    Jensen 29, 38
    Kolmogorov 38
    Markov 38
    probabilistic 38
inferences 159
interquartile range 13
interval
    prediction 161
    tolerance 50, 51, 125
introduction 1
invariance property 139

J

Jacobian 41
Jensen inequality 29, 38
joint
    distributions 49
    moment 34, 37
    probability 46

K

Kendall's
    rank correlation coefficient
        165, 166
    Tau 166
known variance ... 128, 143, 144, 148
Kolmogorov
    –Smirnoff test 167, 168
    –Smirnoff values 170, 171
    inequality 38
Kruskal–Wallis test 173, 174
kurtosis 16

L

Laplace distribution 89
large numbers 45
law
    of exponents 202
    of large numbers 45
    of total probability 26
least squares 157
    estimates 157, 158, 164
    principle 164
lemma Neyman–Pearson 151
likelihood
    function 138, 151
    ratio tests 151
linear combinations of RVs 37
linear regression 157
    multiple 164
linearity of regression 161
location parameter 99, 102
logarithmic function 207
logistic distribution 89, 102
lognormal distribution 100
    definition 97

M

Mann–Whitney
    –Wilcoxon procedure 145
    test 192, 193, 196
Marcienkiewicz's theorem 32, 131
marginal distributions 34
marginal totals 154
Markov inequality 38
mass function 27
matrix
    design 164
    determinant 42

matrix (*continued*)
    positive semi-definite 101
maximum likelihood 138
    estimator 137, 139
mean 30, 45
    arithmetic 7, 8
    deviation 11
    distribution 45
    geometric 8
    harmonic 8
    response 160
    square
        error 135
median 9, 10, 49, 145
    confidence interval 144, 145
    test 185, 192, 194, 200
method of moments 138
midrange 50
minimum variance unbiased
    estimator 136
modal class 9
mode 9
modified Bessel function 104
moments 15, 30, 34
    about the mean 15, 30
    about the origin 15, 30
    factorial 30, 31, 44
    generating function .. 30, 31, 36,
        43, 44
    joint 34, 37
    method of 138
    order 34
most powerful test 151
multinomial coefficient 22
multinomial distribution 70
multiple linear regression 164
multiplication rule 26
multivariate distributions ... 32, 101,
    131

N

natural numbers 201
negative binomial
    distribution 67, 71
    definition 70
    tables 71
negative skew 9
Neyman–Pearson lemma 151

noncentral
    chi-square distribution 100
    t distribution 113
nonparametric statistics 165
normal
    distribution .. 40, 41, 45, 47, 48,
        51, 54, 61, 72, 79, 97, 105,
        110, 112, 113, 115, 131,
        132, 139, 145, 146, 152,
        157, 158, 160–162, 164,
        175, 176, 185, 187, 192,
        193, 200, 201
        definition 99
        multivariate 42, 101, 131
    equations 157, 164
notation 212, 213
n-th root function 202
null hypothesis 147
number
    of failures 70
    of runs 175
    of successes 60, 68, 72
numbers
    natural 201
    real 201
    rencontres 23
    Stirling 22
numerical summary measures 7

O

odds 25
ogive 6
one
    -sample
        Kolmogorov–Smirnoff test
            167
        Z test 128
    -sided
        Chebyshev inequality 38
        confidence interval 141
        test 170
        tolerance interval 126
operating characteristic curves ... 128
order
    counts 21
    of a moment 34
    statistics ... 48–51, 145, 167, 168
        normal distribution 51
        uniform distribution 50

outlier 13, 155

P

p-value 148
parameter
    location 99, 102
    scale 97, 99
    shape 87, 97, 102
Pareto distribution 89
    definition 102
partitions 22
Pascal distribution 67
Pearson's coefficient 16
Pearson's moment 16
percentage points 53
percentiles 11
permutations 20–22, 210
pie chart 6
plots
    box 13
    stem-and-leaf 3
Pochhammer's symbol 206
    definition 204
Poisson
    distribution .. 61, 69, 71, 82, 139
        definition 72
        tables 72
population correction factor 46
positive
    semi-definite matrix 101
    skew 9
power 151
    function distribution 78, 89,
        102
    of a test 147
    sums 208
prediction interval 161
principle least squares 164
probability 19
    axioms 24
    conditional 25
    conservation 31
    density function 28
    histogram 5
    joint 46
    mass function 27
    theorems 24
probable error 12
product rule 19, 20

property, invariance 139

Q

quantile plots 13
quartile 10
    coefficient 16
    deviation 13

R

random
    function 28
    sequence 28
    sums 44
    variables
        continuous 28
        definition 27
        discrete 27
        functions.................. 39
        independent 29, 34, 36
        linear combination 37
        sums 43, 44
randomized block design test 165
range 13, 50
    interquartile 13
    studentized.................. 53
rank 173
    –sum test 192
    correlation coefficient .. 186, 187
    signed 191, 192, 200
rare events 72
ratio
    distribution 41
    tests 151
    variance.................... 173
Rayleigh
    distribution 43, 81, 88, 100, 109
    definition 103
real numbers 201
rectangular distribution definition 75
recursion 205
references.... 131, 156, 172, 174, 191, 200
regression
    analysis 157
    coefficients 159
    linear 157, 161
    multiple linear.............. 164

rejection region 147, 151
relative
    efficiency 135
    frequency.................... 23
    standing..................... 15
rencontres numbers............... 23
repeated trials 21, 59
residuals 158, 164
root mean square................. 13
row total 153
runs test 175

S

sample
    coefficient of determination 159
    correlation coefficient....... 162
    mean....................... 45
    selection.................... 21
    size 51, 142, 148, 165
    space....................... 23
    total 45
    variance.................... 45
sampling distribution............. 44
scale parameter 97, 99
score........................... 165
semi-invariants 32
series 208
shape parameter 87, 97, 102
Sheppard's corrections............ 17
sign test 185
significance..................... 154
    level 147
significant...................... 147
signum function 102
simple event..................... 23
size of sample.......... 51, 142, 148
skewness...................... 9, 16
Smirnoff test 167, 168
soda pop data.................... 1
Spearman's rank correlation
    coefficient......... 186, 187
special functions 201
standard
    deviation 12
    definition 29
    error 12
    order statistics 51
    score 15

standardized range 53
statistics
    complete 136
    distribution-free 165
    nonparametric 165
    order 145, 167, 168
    test 147
stem-and-leaf plot 3
step function 167, 168
Stirling
    formula 205
    numbers 22
stochastic process 28
studentized range distribution 53
success 59, 70
    number 60, 68, 72
sufficient estimator 135
sum of squares 159
    lack of fit 162
    pure error 162
summarizing data 3
sums
    of powers 208
    of random variables 43, 44
swimming pool data 1

T

t distribution . . . 47, 81, 91, 110, 112,
    113, 133, 159–161
    definition 104
tables
    binomial distribution 61
    combinations 210
    contingency 153
    contingency, 2×2 154
    factorials 204
    geometric distribution 67
    Kendall's rank correlation . . 166
    Kolmogorov–Smirnoff . . 170, 171
    Kruskal–Wallis test 174
    Mann–Whiteny U test 193
    negative binomial distribution
        71
    order statistics 51
    permutations 210
    Poisson 72
    runs test 175
    sign test 185

tables (*continued*)
    Spearman's rank correlation
        187
    studentized range 53
    sums of powers 209
    tolerance intervals 127
tabular procedures 3
tail 79, 90, 96, 103, 109, 168, 169
test
    block design 165
    Friedman 165
    goodness of fit 151
    hypothesis . . . 147, 159–162, 165,
        173, 175, 185, 187, 191,
        192, 200
    Kolmogorov–Smirnoff . . 167, 168
    Kruskal–Wallis 173, 174
    likelihood ratio 151
    linearity of regression 161
    Mann–Whitney 192, 196
    Mann–Whitney U 193
    median 185, 192, 194, 200
    most powerful 151
    nonparametric 165
    one-sample Z 128
    power 147
    randomized block design 165
    rank–sum 192
    runs 175
    sign 185
    signed-rank 191, 192, 200
    significance 154
    Smirnoff 167, 168
    statistic 147
    Wallis 173, 174
    Whitney 192, 193
    Whitney test 196
    Wilcoxon 193, 196
    Wilcoxon matched-pairs . . . 191,
        192
    Wilcoxon rank–sum 192
    Wilcoxon signed-rank . . 192, 200
    Z 128
theorems 29–31, 36, 37, 43, 137
    Bayes' 26
    central limit 45
    chi-square distribution 47
    F distribution 47

theorems (*continued*)
    Marcienkiewicz 32, 131
    probability 24
    t distribution 47
ticket data 1
time waiting 66, 72
tolerance
    factors 125
    intervals 50, 51, 125, 126
    one-sided 126
total 45
    weight 8
transformations 40, 41
trials 59, 60, 66, 68, 70
triangular
    distribution 109
    definition 107
trimmed mean 9
two
    -sample Kolmogorov–Smirnoff
        test 168
    -sided test 170
    sample Z test 128
type I error 147
type II error 147, 165

U

unbiased estimator 135, 136
ungrouped data 7
uniform distribution .. 41, 43, 50, 75,
    78, 89, 113, 139
    definition 108
unknown variance 143, 144, 148
URL 1

V

variance 11, 29, 30, 36, 45
    definition 29
    known 128, 143, 144, 148
    minimum 136
    ratio 173
    unknown 143, 144, 148
vector response 164

W

waiting time 66, 72
Wallis test 173, 174
Weibull distribution 88, 109
weighted mean 8

Whitney test 192, 196
Wilcoxon
    matched-pairs 191, 192
    one-sample statistic 144
    rank–sum test 192
    signed-rank test 200
    test 193, 196

Z

Z score 15
Z test 128